S. M. Fitton

The Four Seasons

A Short Account of the Structure of Plants

S. M. Fitton

The Four Seasons
A Short Account of the Structure of Plants

ISBN/EAN: 9783743402294

Manufactured in Europe, USA, Canada, Australia, Japa

Cover: Foto ©berggeist007 / pixelio.de

Manufactured and distributed by brebook publishing software
(www.brebook.com)

S. M. Fitton

The Four Seasons

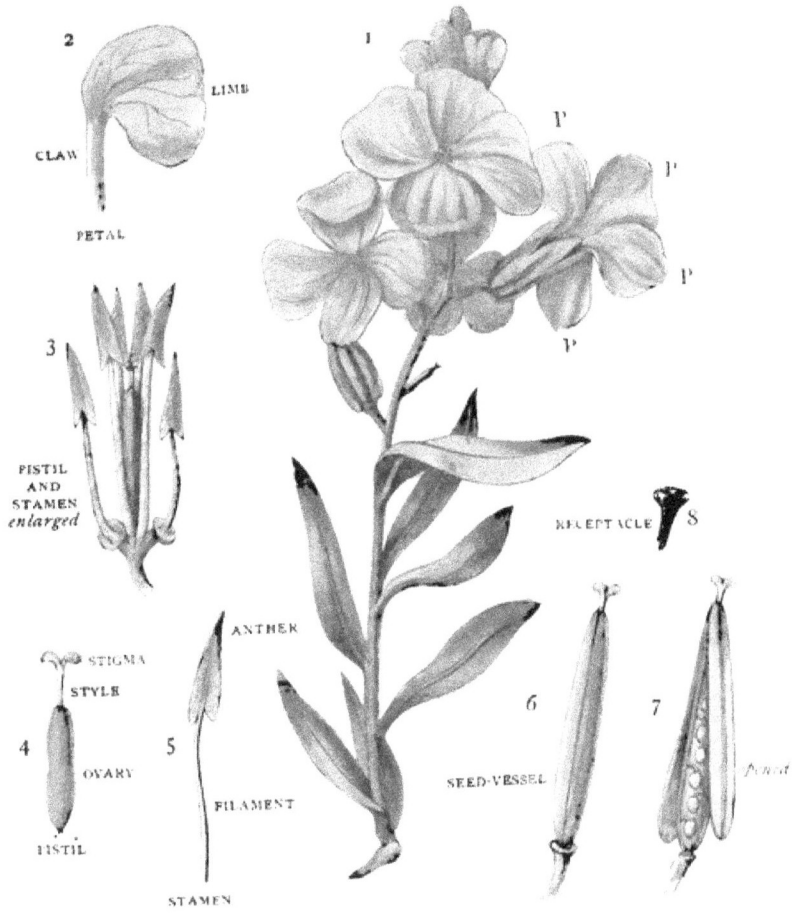

2   LIMB   CLAW   PETAL

1   P   P   P   P

3   PISTIL AND STAMEN *enlarged*

RECEPTACLE 8

STIGMA   STYLE   4   OVARY   PISTIL

ANTHER   5   FILAMENT   STAMEN

6   SEED-VESSEL   7

WALL-FLOWER

THE

Four Seasons;

A SHORT ACCOUNT OF

THE STRUCTURE OF PLANTS.

BEING

Four Lectures,

WRITTEN FOR THE WORKING MEN'S INSTITUTE IN PARIS.

WITH ILLUSTRATIONS.

"O ALL YE GREEN THINGS UPON THE EARTH, BLESS YE THE LORD : PRAISE
HIM AND MAGNIFY HIM FOR EVER."

LONDON:

GRIFFITH AND FARRAN,

SUCCESSORS TO NEWBERY AND HARRIS,

CORNER OF ST. PAUL'S CHURCHYARD.

MDCCCLXV.

TO

SIR WILLIAM JACKSON HOOKER,

D.C.L. OXON. F.R.S. F.S.A. F.L.S.

CORRESPONDING MEMBER OF THE IMPERIAL INSTITUTE OF FRANCE,

AND DIRECTOR OF THE ROYAL BOTANIC GARDENS OF KEW,

MY EXCELLENT OLD FRIEND,

I Dedicate this Volume,

WITH HIS KIND PERMISSION.

S. M. F

Rue de la Ville l'Evêque, Paris,
November, 1864.

CONTENTS.

SPRING.

SUMMER.

AUTUMN.

WINTER.

FOUR SEASONS.

SPRING.

CRU'CIFORM PLANTS.

WALL-FLOWER.   CARDAM'INÉ.

LET us picture to ourselves two persons looking at a water-mill in full work, one of them totally unacquainted with mechanics, the other a civil engineer. How different will be the effect produced upon the mind of each. The one may be impressed with a sense of weariness from the mono-

B

tonous movement of the huge wheels and the plash
of water recurring at regular intervals; while the
other sees an important result produced by ingenious mechanical combinations. And so, in every
subject, an initiated observer sees much that is
hidden from the sight of others. We all instinc-
tively delight in flowers for the sake of their
beauty and sweet smell, and who does not recollect
the gladness of some sick friend when they carried
to his bedside a handful of freshly-gathered violets
in the early Spring? But as soon as we have
seen some of the wonderful contrivances that exist
in even the most simple plant, we look at flowers,
as it were, with new eyes, we read them like a
living book, and thoughts of a still higher perfection fill our hearts. It is with the hope of adding
to your enjoyment during the coming days of
summer, that I now purpose to give you a short
account of the structure and qualities of a few of
the plants that most of us have loved from child-
hood, when we "toddled o'er the brae and pu'd
the gowans fine."

As it will not be possible to show you through
a microscope the internal organization of plants,
I can allude to it but very slightly; and my

descriptions must be limited to their outward framework only, to that which meets the eye, and which all who run may read. You, on your part, will remember that my object is not so much to impart botanical knowledge as to open up to you a never-failing source of innocent and healthful occupation. I have chosen the common Wall-flower for our example at this season, because it is very easily obtained, and it remains in blow for a long time, from the beginning of February usually to the end of May.

You need not be reminded that the part of a plant that grows downwards into the ground is commonly called the Root. Roots do not produce either leaves, buds, or flowers. The branches of the root are called Fibres. From the top of the root the Stem shoots up into the air and light, and is generally clothed with leaves and branches. The points of the stem from which the leaves and branches spring are called the Nodes, or knots. The branch always proceeds from the axil of a leaf, that is to say, from that part of the stem from which a leaf springs. The little stalks that support the leaves in a great number of plants are called Leaf-stalks or Pedicles, from *pedic'ulus*, diminutive of *pes*, the Latin for *foot*, to signify *little foot*,—and the broader spreading part is called the Blade. When a leaf is spoken of the blade is usually meant, whether there be a leaf-stalk or not. On those branches of the Wall-flower that bear flowers the leaf-stalk is hardly perceptible, but in the leaves that grow nearer the ground it is very distinct. The leaf-stalk is composed of several thread-like hollow vessels, which branch out upon the blade as it widens. Some of them run directly to the highest point of the leaf, and form what is called the Midrib, which we see in the middle of most leaves, somewhat depressed on the upper surface, and standing out in relief on the under surface. The smaller ribs that form a sort of network on each side of the midrib are called Nerves or Veins. The word nerve comes from the Greek word νεῦρον, which means *cord, thread*, or *fibre*. In some leaves, those of reeds and grass for instance, the veins do not form a network, but run at once to the highest point, where they meet. They are then said to be Parallel, for want of a more correct expression. The fibres of the root imbibe juices from the earth, which ascend very rapidly through little pipe-like vessels within the stem till they reach the veins of the leaves, where a certain portion is converted into the Sap with which the plant is nourished.

The relative position of the root and the stem is evidently essential to the well-being of the plant; for, when a tree has been turned upside down, root-like fibres strike downwards into the earth from the branches then upon the ground, and young leaves shoot upwards from the root which is exposed to the air and light. In like manner when a branch is inverted each leaf will gradually right itself and return to its natural position with the upper surface towards the sky, and the under surface towards the ground.

Again, if we place a green leaf with its upper surface downwards upon a saucer of water it will keep fresh for some days, but if we place another leaf of the same plant with its under surface upon the water it will keep fresh for as many weeks. The leaves of the mulberry tree show this very clearly. The reason for the difference arises from the different structure of the two surfaces of leaves; the under surface being furnished with a great number of minute vessels, like pores or tubes, through which moisture is imbibed and evaporated, while the upper surface has no pores, and acts as a defence to the leaf. Those breathing pores of leaves have been, not inaptly, likened to the lungs of animals, for they appear to perform very similar functions.

Let us now look at our Wall-flower. (Plate I. Fig. 1.) When real flowers are not to be had we must content ourselves with drawings. The four green leaves tinged with purply-brown, close to the top of the little stalk that supports the flower, are called Se'pals. The word sepal comes from the Latin word *separ*, and signifies *divided into segments*. They form the Ca'lyx or flower-cup. The word calyx comes from the Greek word κάλυξ, which signifies the *bark* or *outer case of flowers*, and not from κύλιξ, which means a *goblet* or *cup*. (Fig. 1.) The calyx generally covers up and protects the young flower while it is not yet unfolded, and afterwards supports the several parts of it in their proper places. The shape of the calyx varies in different plants; in some it is composed of but one piece, when it is said to be Mono-sepalous, and in others it is not required, and therefore does not exist.

The four yellow leaves that come up out of the calyx are called Pet'als, or blossom-leaves. The word petal comes from the Greek word πέταλον, which means a *leaf*. (Fig. 1.) They form the Blossom or Corolla. Corolla is the diminutive of *corona, crown*. The lower whitish part of a petal shaped like that of the Wall-flower, is called the Claw, the spreading upper part the Limb or Border. (Fig. 2.) The claw of the petal is really the leaf-stalk, and the border the blade. In some plants the blossom is Mono-pet'alous, that is to say, it is composed of but one piece : in others there are no petals at all, because they are not wanted.

If we gently pull off the calyx and petals of one of our flowers, we shall find seven threads standing upright on the top of the little stalk. (Fig. 3.) The shortest thread, that which is in the middle, is thicker than the other six, and its lower part is of a pale green colour. It is called the Pistil, and is composed of three parts, the Ovary, the Style, and the Stigma or Summit. (Fig. 4.) The ovary is the lowest and thickest part. It contains the future seeds of the plant. When they grow within it, it becomes very much larger and is then called the Seed-vessel. The style is like a neck or very small column on the top of the ovary, and on the top of the style we find the stigma. The six whitish threads, with yellow heads, that stand round the pistil, are called Stamens. The word stamen comes from the Greek word στήμων, which means a *thread* or *filament*. The word Pistil comes from *pistillum*, the Latin for *pillar*. In several plants the pistil has the form of a pillar. Style comes from the Greek word στύλος, which signifies a *column*. The pistil may be compared to a pillar, of which the ovary represents the pedestal, the style the column, and the stigma or summit the capital. The word *stigma* is the Greek for *seal* or *stamp*. In some plants the stigma resembles the sort of seal or stamp which the ancient Greeks used when they stigmatized or condemned their criminal prisoners. Each stamen (Fig. 5.) is composed of two parts; the whitish thread called a Filament, which is in reality the leaf-stalk, and the head, which is called an Anther, and comes from ἄνθηρος, the Greek for *flowered*, and which corresponds to the blade of the leaf in a modified form. The anther is a sort of bag which is divided into two pouches or cells by a partition down the middle. When the flower unfolds, a fine yellow powder, called Pollen or Farina, is scattered from those cells through an opening or slit. The use of the pollen is to make the seeds within the pistil grow. In the Tulip the pollen is black, and in some

other flowers it is of a bright orange colour, but it is
generally yellow, as we see it in the Wall-flower.
Two of the stamens in the Wall-flower, those
which stand within the side sepals of the calyx, are
shorter than the four that stand between them, their
filaments being bulged or pushed outwards by a
green fleshy substance, or gland, that surrounds each
of them at the base like a ring. You will not be
able to see those glands distinctly without the help
of a magnifying glass. (Fig. 3.)

When the petals of the Wall-flower fall off, the
stamens begin to wither, and the ovary, or seed-
vessel, as it is then called, becomes a long narrow
Pod. (Fig. 6.) The two outer valves or shells of the
pod are separated by a partition, composed of two
very thin gauze-like membranes, which are held together at the edges by some of the thread-like vessels
carried on from the little stalk that supported the
flower. The flat seeds grow upon each surface of
that partition, and their little bent stalks spring alternately from the opposite edges, so that they cannot
crowd against each other. (Fig. 7.) In all plants,
the part that contains the seed is called the Fruit,
whether it be good for us to eat or not and, for this
reason, the little stalk, which at first supports the

flower and then the seed-vessel, is called the Fruit-stalk. The top part of the fruit-stalk is generally somewhat thickened, and upon it all the parts of the flower grow. It is called the Receptacle. (Fig. 8.) The receptacle of the Wall-flower is not conspicuous, but in some other plants it is very large. In the Artichoke, for instance, the receptacle is the part that remains after the leaves and the bristly choke have been removed.

The Wall-flower is Perennial, which means that it lives for more than two years. When we say that a plant lives, nothing more is meant than that it is capable of deriving nourishment from surrounding substances. Plants are said to be Annual when they bear leaves and flowers in one year and then die, and Biennial if they live two years. Perennials of warm regions often live but one year with us, their roots being killed by the cold; our perennials, on the contrary, generally die within a year if they are removed to a hot country, because their roots are deprived of necessary moisture.

Owing to some admirable peculiarity in its structure, each kind of plant is especially adapted to that very climate in which it is destined to live, and to no other, with the exception of grass, which is endowed with the faculty of growing in every climate and in every sort of soil. Some plants require extreme heat, while others grow best in the midst of ice. What are called Alpine plants grow wild on mountains only; other plants of a different nature frequent shady valleys, and some thrive best on open plains. Aquatic plants grow nowhere but in fresh water, and Marine plants are entirely confined to the sea, while some few live on the sea shore, but are never covered by the salt water. The Samphire that is pickled and used in cooking, is one of those coast plants, and the knowledge of its habits afforded unspeakable comfort to some sailors who were once wrecked very near land at Beachy Head in Sussex. The night was dark, and they groped their way up to the top of some rocks at the foot of the cliff, but they expected every moment to be washed away by the high waves. One of them felt a tuft of Samphire that was growing at his feet, and he was then able to assure his companions of comparative safety, which encouraged them to wait without anxiety till daylight, when they could easily attract the attention of the people, and obtain all the help they wanted.

In the Polar regions, where the cold is about 5°

Fahrenheit, or 27° below freezing, plants are low in stature, and have small compressed leaves. The Fir tree too, which thrives best in cold countries, has narrow sharp-pointed leaves that do not obstruct the snow on its way from the sky to the ground. If the leaves were broader the snow might lodge amongst them and break the branches by its weight. The climate becomes hotter and hotter as we descend from the North Pole to the Equator, where the average annual temperature is about 84° Fahrenheit. There the plants require to be shaded from the direct rays of the sun, and accordingly their leaves are generally very broad. The gigantic Palms that grow on the hot tropical plains, and which sometimes attain the height of more than a hundred feet, would suffer from exposure to the great heat were they not protected by proportionately large leaves, which spring out from the top of their tall columnar stems. These fan the air, and afford shelter to the plant on every side, somewhat in the manner of an enormous feathery umbrella. Even in our own little Strawberry plants, we see how carefully the large leaves cover up and protect the flowers and young fruit, keeping them cool and fresh ; but Apple trees and other trees, of which the fruit is less tender and needs no such protection, have comparatively small leaves.

It is to be remarked that the leaves of trees always grow more abundantly on the side of the stem which is exposed to the south, than on the side turned towards the north. The knowledge of this circumstance is invaluable to the poor Laplanders, who might easily lose their way in their long journeys through wild pathless regions if they had not some such natural compass to guide them. The inhabitants of other countries also make use of their observation of the natural appearances that surround them. Some tribes of American Indians, for instance, plant out their corn when the wild Plum blooms, or when the young leaves of the Oak are about the size of a squirrel's ears. Their names, too, of the different months express the state of vegetation, such as the Budding-month, the Flowering-month, the Strawberry-month, the Mulberry-month and so on; and their word that denotes Autumn signifies The fall of the leaf.

The leaves of other plants, as well as those of trees, evince a very decided partiality for the light of the sun, and turn towards it with avidity, as we can see in the pots of Geraniums most people are so

fond of cultivating in their windows. And here it
may be well to observe, that the keeping either
growing plants or cut flowers in sleeping-rooms, is
considered by many persons to be not only unwhole-
some, but dangerous on account of the carbonic acid
gas which they give out, and which is of the same
nature as the fumes of charcoal.

When deprived of sun-light, plants become colourless, and their stems are apt to trail towards the
ground. The outer leaves of a Cabbage are green,
but the short leaves that form the heart are of a
pale yellowish hue. In like manner that part of a
head of Celery which has been what we call "earthed
up" is perfectly white, and the natural dark green
colour melts into the pale leaf-stalks just where the
light begins to approach it. No doubt the warm
light enjoyed by tropical flowers has an effect upon
their colours which are much more glowing and
vivid than the blue, white, and pale lilac so pre-
valent in the flowers of cold countries.

But we must go back to our Wall-flower. The
blossom, as we have seen, is composed of four petals.
(Fig. 1.) They are so placed as to form a sort of
cross, two and two opposite to each other. All
flowers of which the petals are placed in a like man-

ner, and also the plants that bear them, are said to be Cross-shaped or Cruciform. Cruciform plants have so decidedly a marked unmistakable natural character peculiar to themselves, that they form one of the Natural Families or Orders, to some one of which every known plant is assigned. The plants belonging to each natural family are sub-divided into Tribes, and they again into Genera and Species. A Genus, the singular of the word Genera, comprises plants that agree in certain points of structure, generally of their flowers and fruit, and resemble one another more than any other plants. A Species includes plants that agree as to the generic character, but differ from each other in some other respects, it may be in the leaves, the stem, or the root. For instance, there are a great many sorts of Rose which all belong to one particular genus. We have the Dog Rose, the Moss Rose, the China Rose, and several others, all of the Rose genus. Thus it is that every plant is distinguished by two names, one, the Generic name for the genus to which it belongs, the second to denote some peculiarity that marks the species, and which is called the Specific name. If we were to talk of a Dog Rose, a Moss Rose, and a China Rose to a Turk or a

Spaniard, he would not, perhaps, understand us, nor should we recognise those plants if we heard their Turkish or their Spanish names, but with their botanical names, Rosa Canína, Rosa Muscósa, and Rosa In'dicus, we could always convey to one another that we were speaking of three distinct species of the genus Rosa. Latin and Greek being the most universally understood of all languages, the botanical names of plants are generally derived from either one or the other, while the names in the language of the country in which they grow are kept for home use. For instance, the Ranun'culus, or Crow-foot family, has its name from *Rana*, the Latin for *frog*, which expresses the cold damp situations inhabited by frogs, and in which several genera of that tribe grow. Ranunculus pratensis, from *pratum*, the Latin for a *meadow*, denotes the species of Ranunculus found in a meadow; our Buttercup, in fact, called in French, Bouton d'Or. Again, Ranunculus palustris, from *palus*, the Latin for a *marsh*, indicates the species that grows in marshes; and Ranunculus Asiaticus, the showy garden species which first came to us from Asia. The genus Anemoné, belonging to the Ranunculus tribe, has its name from ἄνεμος, the Greek for *wind*, to express that the plant grows in places exposed to the wind, or that its flowers are in blow at the time of the vernal equinox, when the wind is generally pretty high. Our early spring Hepat'ica is a species of Anem'oné. And here it may be well to remark, that in the names of plants, the E at the end is usually sounded, so that Anem'oné is pronounced An-em'-o-ne, like *nee*, and not An-e-mone like *tone*. The Clematis, the Christmas Rose, Larkspur, and Pæony, all of them garden flowers, also belong to the Ranun'culus family. Sometimes a genus takes the name of its discoverer, or that of some remarkable person. For instance, the genus Brown'ea is so called after Robert Brown, the well known botanist, and the genus Fuch'sia after Fuchs, a German botanist. The Fuchsia is a South American plant of the Evening Primrose family.

Some of our most common and useful vegetables are cruciform plants, such as the Cabbage, a numerous genus, of which the cultivated varieties include several sorts of white and red cabbage with smooth or curled leaves, Brocoli, Kale, Brussels Sprouts, and Cauliflower. The peculiarity in the growth of the cauliflower arises from the flower-branches having been squeezed down in such a manner as to force the sap to remain in the fruit-stalks, which causes them to become thick and fleshy, with white grainy heads. A Variety only differs from a species in colour, size, or smell. The sorts of Cabbage we have mentioned may, therefore, be traced back to their common origin, the wild Cabbage, Brassica campestris, of which the Turnip, Mustard, and Colza or Rape that yields so much oil from its seeds, are three species. Water-cresses, Radishes, and Horse-radish also belong to the Cruciform family. The flowers of all those plants are insignificant to the eye, and have generally either white or lemon-coloured blossoms.

No cruciform plants are poisonous in their nature ; but some of them, when they grow near water, acquire such an acrid disagreeable flavour that they cannot be eaten. Our prettiest garden flowers of this family are the Virginian Stock, with blossoms of every possible tint of lilac ; the white, pink, and purple Candy-tuft, which is distinguished by the two adjoining outside petals being larger than the other two ; and our old friend the Wall-flower.

The most beautiful of all Spring flowers are those which belong to the family of Liliaceous plants, of which we see examples in the Tulip, Lily-of-the-valley, Bluebell, Hyacinth, Tuberose, and Fritillary. In one species of Fritillary, the Crown Imperial, an Eastern plant, the flowers contain a great quantity of honey, but the whole plant is of so poisonous a nature that bees will not go near it. The faculty that is given to most creatures of distinguishing noxious from wholesome plants is a great boon to all, for it enables each to find its appropriate food, and none are left without subsistence, what is best liked by some being distasteful to others.

The garden Hyacinth was first brought into Europe from the Levant. A Dutch florist took such a fancy to the flower that he devoted himself to its culture, in which he spared neither trouble nor expense. For a long time he used to throw away the bulbs that produced double flowers, which he looked upon as monstrosities, and cared for those only with single flowers as being the more natural, and therefore the more deserving of admiration. At last the beauty of a splendid double blue one attracted his notice. He took it into his collection, and raised others from it, which he sold for fabulously high prices, two hundred pounds having been paid for a single bulb sometimes.

The bulb contains within it the bud of the future

c 2

plant, which remains quietly in its winter quarters till Spring returns to call it forth. Fibrous roots then shoot downwards to the earth from the flattened under surface of the bulb, and presently a shining green spot appears at its top. The points of leaves begin to push upwards, and, before long, show us the young flower-buds closely packed together, and of a delicate green colour. Gradually the stalk rises, day by day the little buds expand to the light, the blossoms assume their true colours, and then, we have a lovely flower which may well have enchanted the Dutch florist when he beheld it for the first time.

Flowers that have a greater number of petals than usual are said to be Double. Cultivation and richness of soil often bring this about by causing some of the stamens to be transformed into petals. When all the stamens undergo this change the flower is said to be Full. In order to bring back the plant to its natural state, it should be placed in poorer soil, or, in other words, be put upon low diet, which will, by degrees, starve it down.

The White Lily, the pride of every garden, is not in blow till about the middle of May. The blossom is bell-shaped, and the six white filaments support large gold-coloured anthers. The leaves are flat, with parallel veins, and the stem rises from a bulb, of which some of the scales are the bases of the decayed leaves of the preceding year, and others the undeveloped leaves of the year that is to come. The Orange Lily, the Scarlet Martagon, and the Turn-cap Lilies, do not generally flower till the end of June, or early in July.

In Kamtschatka, the bulbs of a species of Lily, the Lilium pomponium, are cultivated for food, as we grow potatoes. They are also dried, ground to powder and used, like flour, for making bread. Those lily-bulbs form part of the winter store of the mice which inhabit that country, and which not only lay them up at the proper season in their underground magazines, but have the forethought to bring them out to dry in the air on sunny days. The women take advantage of this sagacity to replenish their own stock, but they always take care to leave enough for the mice to carry back again, that so their little friends may not die of hunger.

At first sight we might suppose that the pink Guernsey Lily, the Narcissus, Daffodil, and Snow-drop, which belong to the Amaryllis family, and the

Gladi'ola and Crocus, which are of the Iris family, were all liliaceous plants, for in some respects they do appear to resemble them pretty closely ; but both the Amaryllis and Iris families are distinguished by natural characters of their own. Their peculiarities are detailed in Bentham's botanical works.

During the Spring season, we can seldom take a country walk without coming upon a potato field in blossom. Potatoes were first brought from America by Sir Walter Raleigh about the year 1597. The calyx is mono-sepalous, and the blossom mono-petalous, each with four or five clefts in the border. The leaves are generally of a dusky-green colour. In shape the blossom is like an open shallow vase. The dark-green balls that succeed the white or purple blossoms are the seed-vessels or fruit. When the seeds are sown, they produce a great many varieties, so that, in order to ensure the production of potatoes of the same quality as those of the parent plant, they must be propagated by planting what are called the Eyes, which are undeveloped buds growing on the potato itself, just as buds grow on the stems of plants where the leaves are formed. It is a common mistake to suppose that the potatoes we eat are roots. They are really Tubers or excrescences, more or less succulent, that grow on a kind of underground-stem, from which the fibrous roots of the plant descend into the ground. The tuber of an Orchis is generally called a Knob.

The Potato, the Capsicum, and the Tomato, which we find so useful as food; and the Henbane, the Deadly Nightshade, the Belladonna, and the Tobacco, all known to be poisonous when taken internally, belong to the Solánum family. From its tubers the potato plant is called Solánum tuberósum.

Tobacco, as well as the custom of smoking it and taking it powdered as snuff, was brought into England by Sir Walter Raleigh after his discovery of Virginia, to which he gave its name in honour of Queen Elizabeth. King James the First took such a dislike to the smell of this plant, that he wrote a pamphlet against the habit of smoking, and called it " A Counterblast to Tobacco." The name *tobacco* comes from the island of Tobago, where the plant was first discovered. The Americans used to call it Petun, pronounced *pee-toon*. Tobacco has its generic name, Nicotiana, from Nicot, who happened to be French ambassador to Portugal at the time of the introduction of the plant into Europe, and who is said to have offered the first pinch of snuff taken in France to Catherine of Medicis.

In China, tobacco next to tea is supposed to be the best preservative of health, and is constantly used by people of all classes.

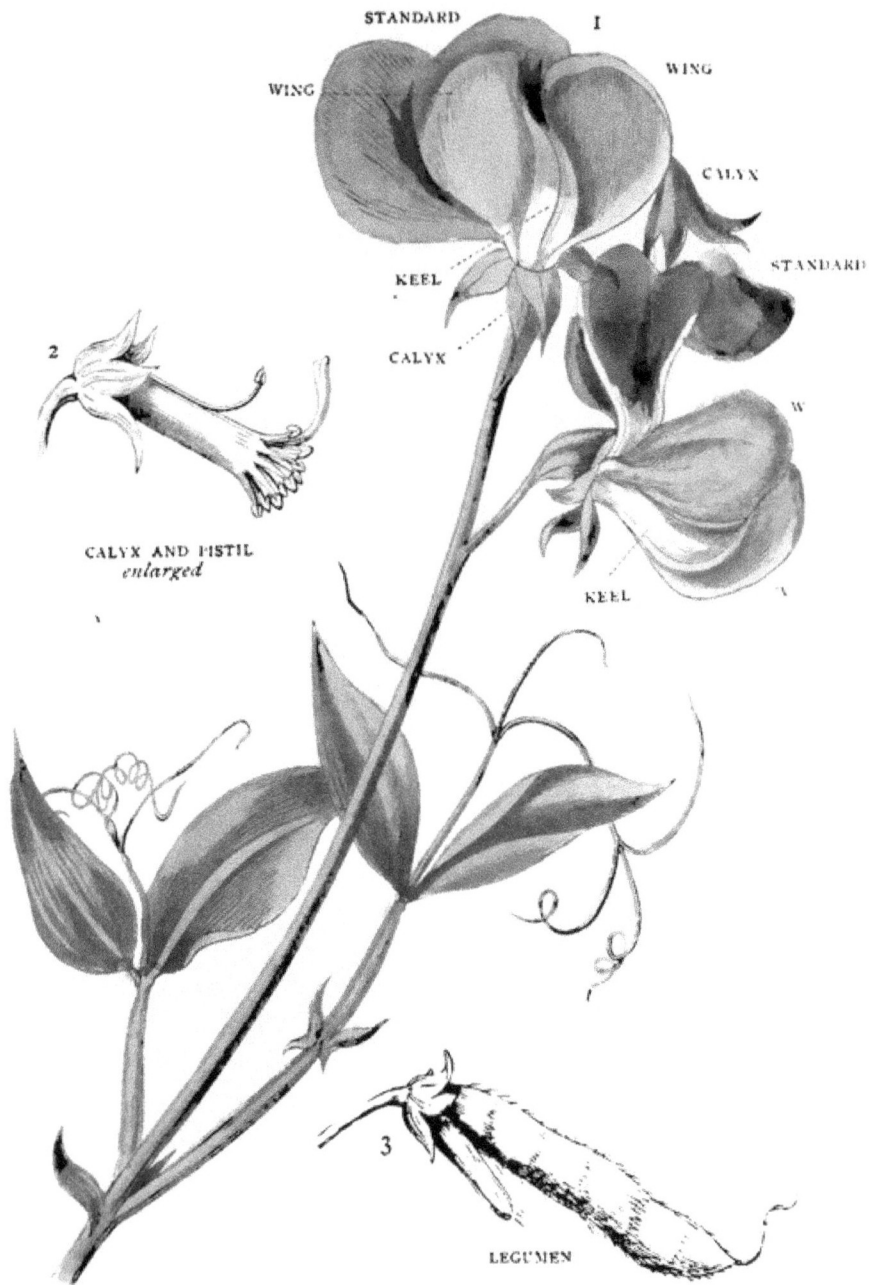

STANDARD   I
WING   WING
                              CALYX
                    STANDARD
KEEL   W
CALYX
                    KEEL

2

CALYX AND PISTIL
enlarged

3

LEGUMEN

SWEET PEA

SUMMER.

LEGÚMINOUS PLANTS.

BROOM.   PEA-FLOWER.

THE garden Sweet Pea is one of the sweetest of
all summer flowers. It belongs to what is called
the Pea-flower tribe, and is a native of Sicily. The
structure of the flowers in that tribe is so peculiar,
that I shall now endeavour to explain it to you.
(Plate II. Fig. 1.) The blossom consists of five petals,
not all of the same shape, for which reason it is said
to be Irregular. When the blossom is composed of
petals all alike in shape, as is the case in the four
petals of the Wall-flower, it is said to be Regular.

The large top petal of the Sweet Pea is called the

Standard, because it stands nearly upright, and its upper edge is somewhat furled back. (Fig. 1.) The standard is much broader than any of the other petals. It is placed outside the other petals and covers them up in the flower-bud. (Fig. 1.) The two petals next below the standard are called the Wings, and sometimes they do look as if they were ready to fly away. They are alike in shape, but reversed in position. (Fig. 1.) The two remaining petals, next below the wings, are more or less united at their lower edges, so as to form a sort of little boat with a curved beak. This is called the Keel. (Fig. 1.) Within the keel we find ten stamens and one pistil. (Fig. 2.) Nine of the filaments are united nearly all the way up, and form a sheath or trough that incloses and protects the future seed-vessel. The tenth filament stands just above the sheath, where its two edges meet. If the sheath were entirely closed like a tube, there would be no room for the young seed-vessel within it to grow ; but this is provided against by the two edges just meeting at first, and afterwards gradually opening as the size of the seed-vessel increases. This we can see with the help of a magnifying-glass and a sharp needle.

The calyx of the Sweet Pea is composed of one piece with five sharp-pointed divisions in the border, the middle segment being always beneath the keel. (Fig. 2.) The two upper segments rest upon the lower part of the standard, and the other two upon the upper edges of the wings.

We already know that the four petals which compose the blossom of the Wall-flower are so placed as to form a cross, and that all plants, of which the blossom has the natural character of being cross-shaped, belong to the natural family of Cruciform or Cruciferous plants. In like manner, the blossom of the Sweet Pea has a distinct natural character of its own, which consists of the peculiar form and position of its five petals. Plants of the Pea-flower Tribe are said to be Papilionáceous, from the fanciful resemblance of the blossom to a butterfly, *papilio* being the Latin for *butterfly*.

Papilionaceous plants belong to the natural family of Legúminous, or Legúmen-bearing plants, the seed-vessel being a sort of pouch, called a Legúmen, which, in appearance, is not unlike a pod. In the Wall-flower, the seed-vessel is a pod, of which the two valves or shells are separated by a partition with the flat seeds growing upon its two surfaces;

but in the legúmen there is no partition, and the
roundish seeds grow alternately on the upper seam
of each shell, as these three little drawings will
show you.

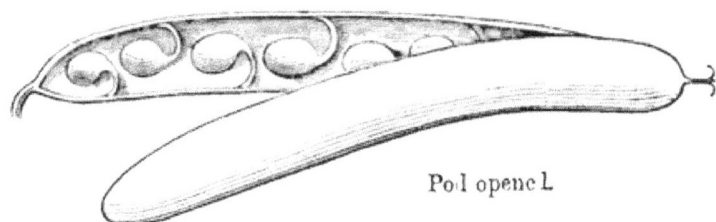

Pod opened.

Legúmen.   Legúmen opened.

The Tendrils or claspers at the end of the leaf-
stalks of the Sweet Pea are merely continuations of
those stalks, which would have been mid-ribs of other
leaflets, had there been any. In some plants the
tendrils turn from left to right, as in the Honey-
suckle for instance ; others, on the contrary, go from
right to left, of which the great Bindweed is an

example; and in the Black Bryony the tendril twists itself a certain number of times one way, and then takes a contrary direction, as if for the purpose of taking a more secure hold. In the Ivy, a number of small fibres grow along the stem and on the branches, on the side next the supporter, to which they fix themselves, like real roots, if it be a tree, or else they cling to naked walls, being covered with hairs which exude a gluey substance that fastens them to smooth surfaces. Those fibres are really tendrils.

The seeds of several of the plants of the Pea tribe are very nutritious, or at least wholesome, as we find in our well known vegetables—Beans, Peas, Lentils, and French Beans or Haricots; and not only the seeds, but the stems and leaves also of Trefoils or Clover, Saintfoin, Lucern, and Vetches, afford excellent fodder for cattle, either in their green state or dried as hay. At first sight, the purple, white, or cream-coloured heads of Clover might be mistaken for so many distinct flowers; but each head consists of a great number of very small flowers, every one of which is composed of its own little calyx, standard, wings, and keel, in most respects like a miniature flower of our Sweet Pea.

The Bean that is cultivated in kitchen-gardens is an Egyptian species of Vetch. Its large seeds, and the seeds of the French Bean also, when quite ripe and split open, sometimes show the structure of the seeds of leguminous plants. In the greater number of flowering plants the seed is composed of three principal parts, the Skin, the leaf-like lobes called Cotyledons, and the future plant or Embryo.

1. The skin, or husk, is merely a sort of case that contains the other parts within it, and defends them from damp and injury.

2. The cotyledons have their name from the Greek word κοτύλη, a *porringer*, or *saucer-shaped vessel*, which they somewhat resemble in form. They are rather succulent in texture, and contain certain juices, with which they help to nourish the newly formed bud. Springing, as they do, from the sides of the minute stem, they surround and nourish the tender little plant, take care of it till it become strong enough to grow of itself, and finally accompany it to the surface, or else, having finished their appointed work, they quietly fade away.

3. The embryo is the future plant in miniature, and is extremely small at first. Linnæus called it the Cor'culum, or little heart. It is that portion of

the seed which all the other parts of it are intended
to protect and nourish. It consists of the Radicle,
from which the root strikes downwards, and the
Plumula, from which the stem shoots upwards.
The plumula has its name from the little feather-
like hairs that generally appear on the bud.

When the seed has but one lobe, plants are said
to be Mono-cotylédonous, and when two, Di-cotylédonous. In all flowering plants, the minute internal
organization of the seed determines the nature and
structure of the plant that succeeds it ; but as
that organization invariably corresponds with certain
visible peculiarities in the stem, the leaves, and
other parts of the plant, it is not necessary to
anatomize the seed in order to know the number
of lobes attached to the embryo. In mono-cotylédonous plants the stem is cylindrical, without
branches ; the pith, wood, and bark are disposed,
as it were at random, in an irregular, unsym-
metrical manner, and the leaves have parallel veins.
Parallel veins or nerves do not often send out side
veins, and when they do so, those lesser veins never
touch or cross each other. We can see these peculiarities in Palms, Grasses, Orchises, and some other
families. In di-cotylédonous plants, on the con-

trary, the stem tapers as it grows ; it is furnished on every side with branches ; the pith, wood, and bark are disposed in a perfectly regular and symmetrical manner, and the leaves have netted veins. The greater number of known trees, shrubs, herbaceous plants, and herbs, are di-cotylédonous.

In fine weather, the flowers of some of the Pea tribe spread out their wings as if to admit the sun's rays, and fold them up again at sunset. Linnæus once sowed in his green-house at Upsal, in Sweden, some seeds of a papilionaceous plant which a friend of his had sent him from a distance. Very soon they produced two beautiful flowers. The gardener was absent when Linnæus first perceived them, so he took a lantern in the evening to show them to him when he came home, but, to his great surprise, they were nowhere to be seen. The next morning, there they were again, looking as fresh and beautiful as ever. The gardener said, "Those cannot be the same flowers ; they must have blown since yesterday." But Linnæus was not so easily satisfied. As soon as it was dark he again visited the plants, and, lifting up the leaves one by one, he found the two flowers folded up and so carefully concealed that it was impossible at first sight to discover where they were. This led him to watch other plants of the same tribe, and he found that they all possessed the faculty of closing their flowers, more or less, at night. For want of a better term he called this "The Sleep of Plants."

That the sleep of plants depends upon the absence of light has been proved by removing a papilionaceous plant into a dark place during the day, when it shut up its petals as if it had been night; and again, by directing towards it during the night a stream of powerful artificial light, in which case the blossom unfolded as if to the sun.

Plants that sleep at night in their own country at the other side of the globe, sometimes reverse the habit with us and sleep through the day, but expand their petals again at sunset. Tropical plants too, in our hothouses, often go to sleep and wake up at regular intervals of twelve hours, as if we had the equal day and night to which they or their parent plants had been born. Our own Trefoils or Clover always shut up their little flowers at the approach of rain, and, when the shower is over, wake up again apparently quite refreshed. But if I say much more about sleeping plants, we shall perhaps go to sleep ourselves.

D

There is a papilionaceous plant in the East Indies, a species of Saintfoin, called the Moving plant, because its leaves move without any apparent cause. The leaf is composed of three lesser leaves or leaflets, those at the sides being smaller than that in the middle. The two side leaflets execute a rapid double movement up and down upon themselves, by night as well as by day, while the middle leaflet sleeps and wakes according to the action of the light.

In the Sensitive plant, which belongs to the Mimosa tribe of Leguminous plants, the leaves lower themselves, and, as it were, droop or shrink away at the slightest touch. Even a breath of wind or a passing cloud will produce a like result, but as soon as the disturbing cause ceases to exist, every part of the plant returns to its natural position. Another species of the Mimosa tribe, a native of Brazil, is the large tree that produces the beautiful dark close-grained wood called Jacaranda or Rosewood, in French, Palisandre. The name rosewood is from the delicate scent of roses emitted by the wood when the tree is first cut down.

The Acacia also, belongs to the Mimosa tribe. Some of its species afford excellent timber, and in several the bark contains certain astringent and tonic properties, which cause it to be much used by tanners and dyers. The legumens of one species are of such a soapy nature that the Indians make a sort of decoction of them, which they use for washing, as a substitute for the soapy seeds of Mimosa saponária.

Several plants send up suckers from their roots as we see in our Rose trees, young Lilacs, and other trees. In the Acacia tree the suckers are innumerable and grow with astonishing rapidity. A farmer who lived at Long Island, in North America, planted fourteen acres with Acacia suckers very soon after his marriage, as a provision for his family. Twenty-two years later, he cut down three hundred pounds worth of timber from his Acacia wood, to enable his eldest son to buy a farm and marry. A few years after that, he gave one of his daughters a like portion; then another of his sons wanted money to enable him to settle, and they again had recourse to the Acacias, till all the eight children were provided for, a quick succession of young suckers filling up the gaps left by the trees that had been cut down to supply the wants of the family.

Very few papilionaceous plants are poisonous.

The seeds of one of them, the West Indian Fish Bean or Jamaica Dogwood, when thrown into water where there are fish, have such an effect upon them, that they come up and float upon the surface and allow themselves to be taken out with the hand. The seeds of the Laburnum and of Lupine are also extremely noxious. The butterfly-shaped yellow blossoms of Laburnum are to be seen in most gardens when Lilac is in blow in the spring of the year, and blue, white, and yellow Lupines flower in July. It is said that in Egypt the people who live near the banks of the Nile destroy the hippopotamus or river-horse, which does great mischief to their fields and gardens at night, by placing near his haunts seeds of Lupine, which he greedily devours.

Some very good dyes are obtained from plants of the Pea tribe that grow wild in England. One species is called Dyer's Greenweed, which yields the best yellow dye we have for colouring wool; another is called Woad, and it affords a blue tint, by means of which the yellow can afterwards be made green. But the best blue dye yet known is the Indigo, which is obtained from an East Indian papilionaceous plant, Indigófera tinctoria. The dye is prepared by steeping the small branches and leaves in water, and drying the sediment they deposit, which is, in fact, the indigo. In its prepared state indigo is poisonous, but the plant itself is harmless. The botanical name of the Dyer's Greenweed is Genis'ta tinctoria, and that of the Woad, Isátis tinctoria. About a century before the Norman conquest, an Earl of Anjou, whose name was Fulke, went on a pilgrimage to the Holy Land. As a sign of humility, he wore a sprig of Genista in his cap, and afterwards took the title of Plantagenet (Planta Genista or Genesta) from it, which his descendants retained.

There is one very common papilionaceous plant, the Furze, which grows wild in many parts of Europe, but it does not bear cold well. Linnæus tried to preserve it under cover, through the winter, in Sweden, and took as much care of it as we do of hothouse plants, but without success. When he visited England in the year 1736, he was so much delighted with the golden bloom of the Furze on Bagshot heath, and on other heaths near London, that it is said he fell on his knees to admire it. In the furze, all the stamens are united into a sheath, over the very short and few-seeded pod.

It would hardly be like Summer if I did not say a few words about Roses, which are to be seen in every garden at this season. The most important of the natural characters of the Rose family, is the growth of the stamens within the throat of the calyx, for it indicates that the fruit, which succeeds the flower, is always wholesome, even though poison may lurk in other parts of the plant. The Rose family is generally divided into the Almond, the Rose, and the Apple tribes.

The Almond tribe includes the Plum, the Cherry, and the Apricot, three species of the genus Prúnus; and the Almond, the Peach, and the Nectarine, three species of the genus Amyg'dalus. In the Almond tribe the calyx falls off when the plant has done flowering.

The Rose tribe includes, amongst other genera, the Blackberry, of which the Raspberry is a species, the Strawberry, and the Rose. In the Rose tribe the calyx remains after the plant has done flowering.

The Apple tribe includes the Pyrus or Pear and Apple genus; the Hawthorn, the Medlar, and the Mountain Ash. In the Apple tribe the calyx-tube becomes the seed-vessel or fruit. We can see a little crown of the withered sepals on the top of unripe apples and pears.

In plants of the Rose family the five petals are hollowed into the shape of saucers or little bowls, and are either white, or white tinted with a delicate pink or pale lilac colour. In the Rose genus the calyx is either globular, or oval, or shaped something like a boy's top, the border being divided into five deeply cut sepals. The Dog Rose is a very good example, and is to be found in flower in most hedges all through the Summer.

What we call the thorns of Rose trees are really prickles, and can be peeled off with the bark, while thorns grow from the wood itself and remain after the bark has been removed. Thorns are, in reality, undeveloped buds, which a rich soil and cultivation would convert into branches and leaves. The wood having a tendency to grow upwards, thorns also point either upwards or straight out from the stem; but prickles, proceeding as they do from the bark only, generally point downwards. They are really enlarged hairs.

Roses are found chiefly in temperate or cold climates north of the equator. Asia, which may be called the Land of Roses, produces a greater

number of species than any other part of the globe.
In Egypt, Roses used to be the emblems of silence.
The Egyptian goddess Isis, and her son Harpocrates,
the god of silence, were always represented with
chaplets of Roses on their heads.   Hence, probably,
the expression, "Under the Rose," to imply something told to another in confidence.

During the civil wars in England, in the reign of
Henry the Sixth, a white rose distinguished the
partisans of the house of York, and a red rose those
of the house of Lancaster, for which reason those
wars are often called the Wars of the Roses.

The delightful Eastern perfume called Otto or
Attar of Roses, with which Indian shawls are often
scented, is an essential oil obtained from the petals
of the Damask or Damascus Rose.   In Europe the
rose water used in medicine is distilled from the
petals of the sweetest-smelling roses that can be
obtained.

1

2
FLORET OF
CIRCUMFERENCE

3
FLORET
OF CENTRE

CHINA ASTER

AUTUMN.

Front view.

Back view.

DAISY.

AT first sight, a China Aster appears to be composed of several flat narrow petals, ranged in a circle of diverging rays round a button-like centre consisting of a great number of yellow dots (Plate III. Fig. 1.) But if we pull out all those petals and yellow dots, we shall find that what we thought were flat narrow petals only, are really little flowers or florets, the blossom of each being composed of one long petal, in shape somewhat like a spear with a blunt point. This petal forms a very short tube at the base, through which the pistil rises from the top of a little egg-shaped seed that is crowned with a ring of silky white hairs, called Pappus. (Fig. 2.) There are no stamens in the flowers of the circumference in this plant. What we mistook for yellow dots are, in reality, beautiful little flowers or florets, each blossom being composed of one petal with five divisions in the border, and not unlike a funnel in shape. There are five stamens, and the anthers are united side by side in such a manner as to form a cylinder, through which the pistil rises from the little egg-shaped seed below, crowned, like those of the circumference florets, with fine soft white hairs. (Fig. 3.)

Flowers that consist of a number of florets collected together into a head, and surrounded by a sort of general calyx, the whole having the appearance of a single flower, like our China Aster, are said to be Composite. The plants that bear such flowers form the natural family of Composite plants. The calyx peculiar to this family is called an Involúcrum, a Latin word, from *involvo* to *wrap up*, and which means a *covering* or *case*.

Composite flowers are all so far alike, that the family has a very distinct natural character, as you may see by comparing the China Aster with Marigolds, Sunflowers, and Daisies. You must always take care to gather a Daisy before sunset and on a fine day, for it opens every morning to the rising sun, and shuts itself up at the approach of rain, and every evening when the sun declines, as if to take its rest. Its name Daisy, Day's Eye, or the eye of day, comes from this habit of early rising.

The structure of the involucrum and the disposition of the hairs and chaff-like little scales that accompany the seeds, are of great importance in enabling us to distinguish composite plants from each other amongst themselves, but the union of the anthers into the form of a cylinder is the great distinction of the Composite family, because that does not exist in any other plants. There are some flowers of which the general appearance might naturally lead us to imagine that they were members of this family; the Scabious and Teasal for instance, for they too are composed of a number of small florets collected into a head, which has the appearance of a single flower, but in neither of them does the union of the anthers in the form of a cylinder exist.

In some composite plants the involucrum is composed of a single row of leaves or scales ; in others
there are two rows, and in others again the scales
overlap one another alternately, and in a perfectly
symmetrical manner, as we see them in the common
Artichoke, which is a composite plant. In many
instances the involucrum opens as if to make room
at the right time for the florets to expand, and
then, when the petals have fallen off, it again closes,
apparently in order to protect the young seeds; but
as the seeds ripen and increase in size, it re-opens
to give them space ; and in some plants it turns
entirely back to let them escape. The Coltsfoot and
Dandelion are in this condition when we see their
heads covered with light Down.

The seeds in several of the species are very remarkable ; they are invariably placed below the
blossom, and there is never more than one to each
floret. In several plants they are topped with a
most beautiful down, consisting of a number of
spokes or rays; the spokes themselves, too, are sometimes branched or feathered, and in some plants—in
the Dandelion, for example—an entire crown or
wheel, formed of the branches of the down, is fixed
upon a sort of stem or pillar, which is itself attached

to the seed; something in this manner, only in-
finitely more delicate and light-looking than any
drawing ever can be.

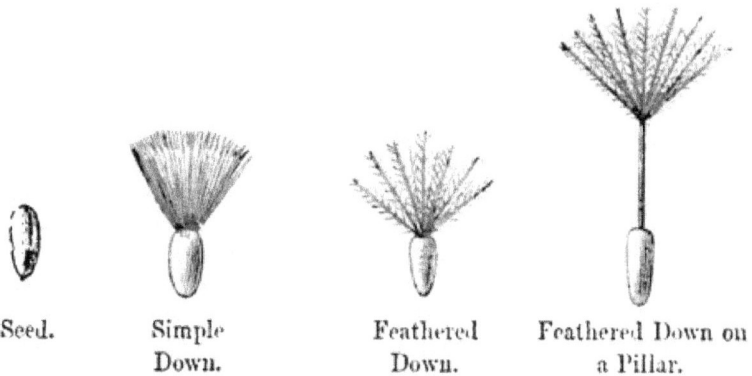

Seed.   Simple   Feathered   Feathered Down on
          Down.   Down.   a Pillar.

The down is a beautiful object when seen through
the microscope, and its use is most important, for it
enables the wind to carry the seeds to very considerable distances from the parent plant, and to
sow them, as it were, in situations which they might
otherwise, probably, have never reached.

It is highly interesting to observe how carefully
the dispersion of the seeds of plants is insured.
Those seeds, for instance, which grow best in some
peculiar soil, such as the seeds of the Arum, are
small in size, and so heavy that they fall straight
down into the ground as soon as the ripe seed-
vessel opens to let them out, and there they grow

without further care, almost on the same spot
where the parent plant flourished before them.
Some larger seeds are yet so very light, that they
could be borne aloft by the wind, were they not
furnished with little grappling-hooks that prevent
them from straying too far before they settle upon
the ground. Others, again, have wings, that so,
when ripe, they may be dispersed in the air before
they are sown. Were they all to sink into the
earth at once, they might come up so close together as to smother many of the future plants.
The seeds of the Ash and Elm are of this description. Some seeds are scattered by being spurted
or jerked out by the parent plant itself. Those
of the Wood Sorrel are thrown off in this manner,
the seed-vessel being so constructed that, when it
begins to dry, it bursts open, and, in a moment, is
violently turned inside out, upon which the seeds
dart away in every direction and to a considerable
distance. When Oats are ripe, the grains are thrown
from the calyx with such a loud sound that, in
passing near an oat-field, if the day be dry and
fine, we may hear the crackling noise.

In Pine or Fir trees the seed-vessel is a Cone with
scales lying over each other like tiles on a roof, and

which have the appearance of being one undivided body. In this state the cone hangs upon the tree during the winter season, and protects the seeds which it contains; but as soon as warm weather comes again, the scales of the cone begin to shrink and separate, leaving openings for the ripe seeds to escape. If a number of cones happen to burst at the same moment, which is often the case, the noise can be heard at some distance.

Birds help to disperse seeds when the seed-vessel forms part of their food, which is the case with Cherries, Sloes, Haws, and other fruits of a like nature. They carry away a Cherry, for instance, till they find some convenient place where they can eat the pulpy seed-vessel at their leisure and, that done, they drop the stone with the seed or kernel in it, down upon the ground.

It sometimes happens that the seed itself is destroyed. In the Strawberry we ourselves eat the seeds along with the pulp, and slugs do the same even before the fruit ripens; but losses of this kind are provided against by the faculty the plant possesses of throwing out fresh shoots or suckers, much in the same manner as the stoloniferous grasses of which I shall presently speak.

Amongst our autumn garden flowers of the Composite family, the most beautiful are the Dahlias from Peru, the French and African Marigolds, and the Cinerarias of every possible hue, from the Cape of Good Hope and the Canary Islands. You will perhaps be surprised to hear that the showy Cinerarias of the greenhouse are only varieties of one species of the genus Senecio, of which the common Groundsel, the delight of Canary birds, is another species. The handsome Globe Thistle, the Scotch or Cotton Thistle and, flowering almost as late as christmas, the brilliant white, pink, and pale yellow Chrysanthemums that look so gay when few other flowers are in blow, are also composite plants. Our garden species of Chrysanthemum, the Chrysanthemum Indicum, came originally from China. In Chinese and Japanese drawings, and upon Indian lacker-work and porcelain, the Chrysanthemum is frequently represented.

In cottage gardens we seldom fail to meet with golden-yellow Marigolds, Sunflowers, Bluebottles or Cornflowers of every shade from white and light blue to deep purple ; white Ox-eye Daisies, deep yellow Corn-marigolds, and Chamomile ; and, if a hive of bees be there, Michaelmas Daisies are rarely absent. The Michaelmas Daisy belongs to a North-American species of the genus Aster, to which the Eastern plant, our China Aster, either belongs, or is very nearly allied.

In the kitchen garden we have the common Artichoke, the Jerusalem Artichoke, which is a species of Sunflower, Helianthus tuberosus; Wormwood or Absinth, Tansy, though now somewhat old-fashioned, Southernwood, and Taragon, the last four being species of the genus Artemisia. Another species of Artemisia, Artemisia Chinensis, yields the Moxa of China, a substance often used in surgery for cauterising or artificial burning. We have Lettuce too, and Salsify, and Chicory. As for weeds, there is no lack of them, for have we not Groundsel and Dandelion? The name Dandelion is the corruption of *dent-de-lion,* which is meant to express the supposed resemblance of the jagged leaves of the plant to the teeth of a lion.

The Composite family is the best defined and the most extensively diffused of all the natural orders. Yellow and lilac are the prevailing colours of the flowers and, though most of the plants are bitter, none of them are poisonous. It is to be remarked, that the colours in Autumn flowers are much more

E

warm and rich in tone than those in our Spring
flowers.   At St. Helena, the composite plants are
trees, and in Chili they are shrubs.   In Europe they
are seldom more than herbs, unless it be in Italy and
the south of France and Spain.

But the plants in which we must always take
the greatest interest are those which give us grass
and corn.   They form the great natural family,
called Graminæ, from *gramina*, the Latin word
for *grass*.   The flowers of the grasses are very
modest and unpretending in their appearance, but
they are not less beautifully constructed than those
of more striking plants.

The care taken by Nature to ensure the production of grass is truly wonderful.   Though the leaves
be trodden down and consumed, the roots still continue to increase and, as the stalks that support the
flowers are seldom eaten by cattle, the seeds are
generally allowed to ripen.   Some of the grasses that
grow on very high mountains, where the heat is not
sufficiently powerful to ripen the seed, are propagated by shoots or suckers, which rise from the root,
and then run along the ground and take root themselves.   Plants of this kind are said to be Stoloniferous, from *stolon*, the Latin for a *sucker of a plant*.

Other grasses are propagated in a manner yet more surprising. The seeds begin to grow while they are within the Husk, as the calyx of grass is called, and young plants are there formed with little leaves and roots of their own ; these fall to the ground, take root, and continue to grow like the parent plant from which they sprang. Grasses such as they are said to be Vivip'arous. In warm countries green turf is unknown, for grasses do not grow there compactly together as in our meadows, but are dispersed either singly or in groups like other plants. Tropical grasses often attain the height of trees ; their leaves are broad, and the flowers soft, downy, and elegant in form. In Europe the best fodder grasses are not higher than three or four feet above the ground. If they grow beyond that height, they are apt to become wiry and hard. In Brazil, though the grasses are gigantic in stature, they are tender, delicate, and very juicy.

The Romans used to bestow a crown made of woven grass as a reward to their generals who, by their skill and courage, had saved an army from destruction ; and, though in itself of so little value, that grassy crown was always eagerly contested even by grave old warriors.

Wheat, Barley, Oats, Rye, and every kind of corn, are the seeds of grasses, and the Sugar-cane, Rice, Maize, and every sort of Reed, including the South American Pampas Grass, belong to the grass family. The fruit of grass is called the Grain. Every kind of grass seed in its healthy state, is wholesome and nutritious.

Some of the Reeds of Brazil are invaluable to the hunters in that country. In remote places, if they happen to be fatigued and thirsty, they have only to cut off a reed below the joint to find the stem beneath it filled with a cool pleasant liquid which is perfectly harmless. And as the reeds form a thorny thicket which wild animals cannot easily penetrate, the hunters can enjoy the refreshment afforded by those living fountains without fear of being molested.

The plants that form what is called the Sedge tribe very nearly resemble grasses, but they are scarcely ever eaten by cattle. The most remarkable of the sedges is the Papyrus, which grows in Sicily and on the banks of the Nile. The ancient Egyptians often represented the Papyrus on their monuments, which shows how highly they valued it. They made vases of the roots, and boats of the stalks woven together like basket-work and then coated over with some sort of resin. When the bark is peeled off the stalks of the plant, the pith that remains can easily be separated into very thin layers, and of those layers they made not only paper but cloth for dresses. Their way of making paper was to place a number of layers of pith close beside each other, with as many more layers ranged across them, and then to wet the whole with water, which made the pieces stick together. When this was pressed and dried, it was equal to our paper in solidity and lightness.

The word *paper* comes from the name of the plant Papyrus, and from the ancient custom of writing upon leaves we say the leaves of a book; but the word *book* is from the Saxon word *Boc*, or Beech, thin tablets of beech-wood having been used by our ancestors for writing upon, as ivory or porcelain tablets are sometimes used to this day. *Liber*, the Latin for *book*, signifies the bark of a tree upon which people also formerly wrote, and *volumen*, a *roll*, was that manuscript rolled up, for bark could not be folded like paper without cracking or splitting. From *liber* and *volumen* come our English words *library* and *volume*. Paper was first made in England about the year 1300, and it appears to have been first made of linen rags half a century later. The first paper-mill in England was erected in the year 1588.

If you wish to become acquainted with other Spring, Summer, and Autumn plants, besides those of which I have spoken, I can safely recommend to your notice Bentham's "Handbook of the British Flora" as an admirable work, and very concisely written.

When we look upon Autumn flowers, Dickens' delightful picture of that season naturally recurs to our thoughts, and I think you may like to be reminded of it, though it has nothing to do with plants :—

"There is no month in the whole year, in which nature wears a more beautiful appearance, than in the month of August. Spring has many beauties, and May is a fresh and blooming month, but the charms of that time of year are enhanced by their contrast with the winter season. August has no such advantage. It comes when we remember nothing but clear skies, green fields, and sweet-smelling flowers—when the recollection of snow, and ice, and bleak winds, has faded from our minds

as completely as they have disappeared from the
earth, — and yet what a pleasant time it is!
Orchards and corn-fields ring with the hum of
labour ; trees bend beneath the thick clusters of
rich fruit which bow their branches to the ground ;
and the corn, piled in graceful sheaves, or waving
in every light breath that sweeps above it, as if it
wooed the sickle, tinges the landscape with a golden
hue. A mellow softness appears to hang over the
whole earth ; the influence of the season seems to
extend itself to the very waggon, whose slow
motion across the well-reaped field, is perceptible
only to the eye, but strikes with no harsh sound on
the ear."

FLOWERLESS PLANTS.

FERN.　　　　　MUSHROOM.　　　　　MOSS.

You have seen that in some plants there are no petals, and in others no calyx, but I have not yet spoken to you of any plants without stamens and pistils, which are, in themselves, sufficient to constitute a perfect flower, for they alone produce the seeds of future plants. In Ferns, Mosses, Lichens—pronounced li-kens, the *i* long—Mushrooms or Funguses, Sea-weeds, and other plants of a similar nature, there is neither stamen nor pistil, nor what can be called any true seeds, and consequently those plants

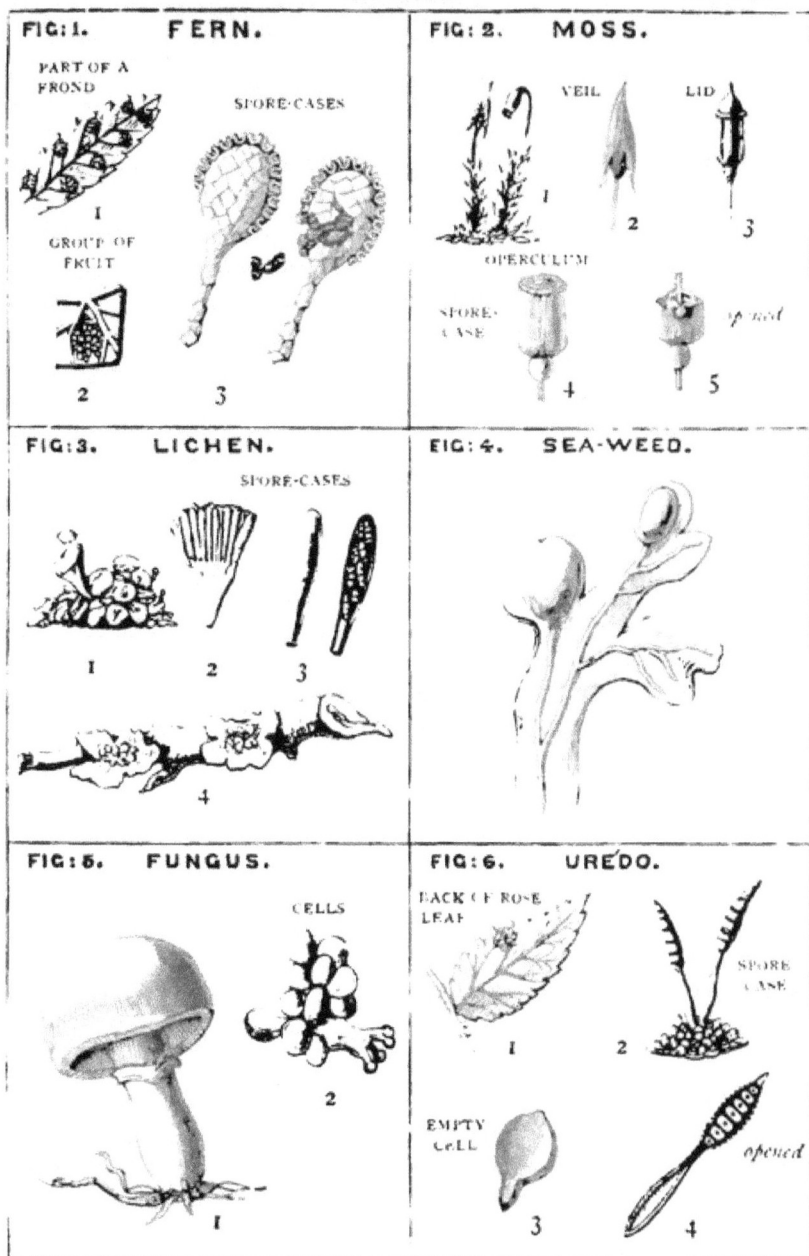

FIG:1.　FERN.

PART OF A FROND

SPORE-CASES

1

GROUP OF FRUIT

2　　3

FIG:2.　MOSS.

VEIL　LID

1

2　　3

OPERCULUM

SPORE-CASE

opened

4　　5

FIG:3.　LICHEN.

SPORE-CASES

1　　2　　3

4

FIG:4.　SEA-WEED.

FIG:5.　FUNGUS.

CELLS

2

1

FIG:6.　UREDO.

BACK OF ROSE LEAF

SPORE CASE

1　　2

EMPTY CELL

opened

3　　4

FLOWERLESS PLANTS

are said to be Flowerless. Their fructification is carried on by means of exceedingly fine dust-like minute grains or granules, called Spores, which are sent forth from the parent plant in a very wonderful manner, and which may possibly be miniature buds of myriads of future plants, totally invisible to our eyes and also to the best microscopes yet invented. But the precise manner in which the fructification does take place is not clearly understood at present.

The visible structure of some flowerless plants might lead us to infer that they were mono-cotyledonous, but that is not the case. In the Fern tribe, the green leaf-like parts that we should naturally feel inclined to call leaves, bear the fruit, as it is called, grouped in round or oblong clusters upon their under surface, or else growing in little cups upon their edges. Those fruit-bearing branches are called Fronds, to distinguish them from what we mean to express when we talk of the leaves of other plants. At the lower part of the stalk of the plant, several blackish root-like fibres shoot out, mixed with membranous scales of a red-brown colour. Those scales protect the young shoots, which at first appear in the form

of a pale-green tuft, surrounded by old shoots of the
preceding year.

Ferns are generally herbaceous with us, but
in tropical regions they become woody, and take
the form and size of trees. They compose about
a tenth part of the plants that grow in those
countries. They are so very beautiful and so
graceful in form, that they are much sought by
amateur gardeners; and few things in vegetation
are more interesting to watch than the gradual
uncurling of the young fronds as they peep up from
the brown stems covered with pointed scales. The
groups of yellow or brown grain-like dots, which
we see on the under surface of some of the fronds,
are composed of spore-cases, which open at the
proper moment, to let the ripe spores escape. Perhaps
this drawing may give you some slight idea of
their general structure. (Plate IV. Fig. 1.)

The fronds of most of the species of Fern, if
cut when fully grown, and then properly dried,
make a thatch more durable than any sort of
straw. The root of one sort, called the Flowering
Fern, because the fructification forms a sort of
panicle or loose bunch at the top of the frond,
when boiled in water, is employed in the north of

Europe like starch, to stiffen linen. The aromatic syrup called Capillaire, is prepared from a British species of Fern, the Adiantum Capillis Veneris. A species found in North America, called the Sensitive Fern, is said to wither immediately if it be touched by the human hand, though it may be touched by anything else without the slightest injury. This experiment was repeated several times, and always with the same result, by Sprengel, a well-known German botanist. You will find every necessary detail respecting this tribe of plants in Moore's excellent "Handbook of British Ferns."

The Mosses have roots and leaves apparently like those of other plants, but their fruit is very different. Small crisp, almost hair-like threads, resembling the filaments of stamens, generally grow out of the bosom of the green leaves, and support little pointed oval heads, that we might at first sight suppose were anthers. Those crisp threads are fruit-stalks, and the heads capsules, which is the name given to the case or box-like seed-vessels. The shape of the capsule varies in different kinds of moss. It is usually almost entirely covered with a sort of veil or cap, called a Calyptra or veil, from καλύπτω, the Greek word for *to veil*.

In form it is not unlike a little extinguisher, and is composed of fine silky hairs. It is of a yellow colour at its base, and of a rich brown at the highest point. When we have carefully removed the veil, we come to a four-sided urn-shaped head with a sort of conical crown, standing on the top of the stalk. On detaching this crown, which is neither so high nor so sharp-pointed as the veil, and is, as it were, hemmed in round its base, we come to a sort of semi-transparent, slightly-convex inner lid stretched over the top of the four-sided urn, something in the manner of the parchment on the head of a drum. This inner lid is called the Oper'culum, which is the Latin for a *lid* or *cover*. It is kept or held in its place by one or two rows of fringe, of great delicacy and surprising regularity as to the number of the teeth that compose it. It is some different peculiarity in the structure of this fringe round the mouth of the capsule which determines the character of the several genera of Mosses. If we raise the operculum, we see that the hollow urn below it encloses four cells or spore-cases, (Fig. 2.) each cell filled with innumerable dust-like granules, of a bright green colour. Those granules are the spores. As soon as the spores are ripe, the capsules bend a little to one side towards the ground upon the crisp thread-like fruit-stalks; the brown veil, the pointed cover, and the inner lid, all fall off, and the dust-like spores are poured out and scattered upon the ground, where they reproduce other plants of the same kind as the parent plant.

The form of the leaves of Mosses is extremely simple. They are all destitute of leaf-stalks, and are never winged or divided. Mosses are generally perennial and evergreen, and are capable of growing in much colder climates and situations than most other vegetables. In the dreary country round Spitzbergen, the rocks which rise out of masses of ice almost as hard as they, are thickly clothed with moss. Crantz, a German botanist who travelled in Greenland, counted more than twenty different sorts without moving from a rock upon which he one day happened to be seated.

Mosses possess the singular faculty of reviving when moistened, after having become dry and to all appearance withered; and even after they have been gathered and kept in a dry state for some years, if put into water, every part of them will expand and become almost as fresh as if they were really growing. They overspread the trunks and roots of trees, and defend them in the winter from frost, and cold biting winds; in wet weather they preserve them from decay—and, during drought, provide them with moisture and protect them from the scorching heat of the sun. It has been observed, too, that moss grows principally on that side of the ·trunks and branches of trees which is exposed to a northern aspect.

The poor Laplanders derive several of their comforts from mosses. Of the Golden Maiden-hair, a very large species, they form excellent beds by cutting thick layers of it, one of which serves as a mattress and another as a coverlet. Linnæus tells us that he often slept in such a bed when he was travelling in Lapland. Those mossy cushions are very elastic, and so light, that a bed may be rolled up into a parcel small enough to be carried by a man under his arm, so that the inhabitants of that region can easily take them about with them in their journeys. Such beds do not grow hard by pressure, and, even when they lose some of their elasticity from long use, it can soon be restored by plunging them in water.

The Lapland women make great use of the grey Bog-Moss, which is particularly soft, like thick fur, or a fleece of wool. They wrap up their infants in it without any other clothing, and place them in cradles made of leather and lined with moss, and in those soft, warm nests, the little creatures are completely defended from the cold. The Greenlanders use this sort of moss for tinder, and for wicks to their oil-lamps.

The structure of the Mosses is so beautiful and so wonderful, that they are objects of the greatest interest and admiration to all who study them. Mungo Park, a Scotchman of undoubted fortitude and courage, who ventured alone into the great unknown regions of Africa, in the account he gives us of his travels, says : "I found myself in the midst of a vast wilderness, helpless and alone, surrounded by savage animals, and by men still more savage. I was at least five hundred miles from the nearest European settlement ; I considered my fate as certain, and that I had no alternative but to lie down and perish. At that moment, the extraordinary beauty of a small moss irresistibly caught my eye, and, though the whole plant was not so large as the top of one of my fingers, I could not contemplate the delicate conformation of its little roots, leaves, and capsules, without feeling intense admiration. Can that Being, I thought, who planted, watered, and brought to perfection in this obscure part of the world, a thing which appears to be of such small importance, look with unconcern upon the loneliness and sufferings of creatures made after His own image? Reflections like these would not allow me to despair. I started up and, regardless of hunger and fatigue, travelled forwards, assured that relief was at hand; and I was not disappointed."

The rough yellow and bluish-green crusts that we find growing upon the stems and branches of old Apple trees and Gooseberry bushes, are called Lichens. (Fig. 3.) Lichens have neither stalks, leaves, nor roots. They generally consist of a sort of flattened frond, which forms an irregular patch or a roundish rosette with the outer edges more or less cut or notched into scollops. The under surface is fixed upon the bark of trees or upon rocks, old buildings, palings, and other solid bodies, by means of certain little black root-like fibres, which are not real roots, but cells prolonged into that fibrous form. The upper surface is usually sprinkled with raised spots, some of them like warts, others like little cups lined with yellow, or edged with scarlet, and others again are still more open, like shallow saucers. In those raised spots the fructification is going on in various stages of its progress towards maturity. On opening some of them vertically, we find in the cavity several cells placed side by side. Each of those cells contains lesser cells, which are supposed to be the spore-cases. (Fig. 3.)

Some Lichens form a scaly, or a foliaceous crust, some consist of nothing but a very fine powder or dust, and others are composed of masses of thread-like substances. Several botanists are of opinion that the light cobweb-like matter that overruns the grass early in the morning, in spring and autumn, is of this nature. They are all supposed to absorb their nourishment from the air. Like the Mosses, Lichens are found to thrive in every climate, and like them they have the faculty of reviving, when placed in situations adapted to them, though they may have been kept in a dry state for many years. They are not destroyed by either excessive heat or extreme cold, and are found growing where no other vegetation is to be seen.

One species of Lichen, called the Reindeer Moss, is the most valuable vegetable that grows in Lapland, for it is the principal food of the reindeer, without which animal the people in that miserable country could scarcely subsist. The reindeer draws them in sledges over tracts buried in snow ; its flesh and milk afford them nourishment ; its skin clothing ; and even its bones and sinews are converted into several useful articles. This Lichen grows to the height of at least a foot, and is of a whitish or very pale grey colour, so that, where it grows, the ground looks as if it were covered with snow. In our own country it is found in some mountainous situations, but seldom attains the height of more than six inches. The inhabitants of Iceland contrive to prepare very nourishing food from another species of Lichen called the Iceland Moss, which grows there abundantly. The plants called Tripe de Roche, upon which the Canadian hunters often subsist when better food is scarce, are different species of eatable Lichens.

Several species of other Lichens afford very beautiful dyes. One of them, the Dyer's Lichen or Orchall, Roccella tinctoria, is especially valuable on account of the fine purple and crimson colour it gives to wool and silk. It comes chiefly from the Canary Islands. The purple powder called Cudbear, a corruption of Cuthbert, the name of its discoverer, which dyes cotton, but no animal substance, is obtained from another species, the Lecanóra perella, which is common in England.

There is a species of Lichen, the Lecanora esculenta, which is looked upon as a great delicacy. It is found on the most barren mountains in the deserts of Tartary and grows upon the ground amongst the rough stones, which it resembles so nearly in colour that it requires an experienced eye to find it. Monsieur Parrot, a French traveller in Persia, relates that early in the year 1828, large quantities of the Lecanora fell like a shower of rain in different parts of that country, no person being able to discover whence it came. The cattle eagerly devoured it, and the poor people collected it as if it had been manna from Heaven, and made it into excellent bread.

The nearer we approach the north and south poles, the more we find the earth to abound with Lichens; the kind of plants next in number, in drawing towards the equator, being the Mosses, and next to them the Grasses.

What we call Sea-weeds, are plants that grow in sea water. They include several genera. Other plants thrive only in fresh water, and include several species of Conferva and other genera. Both kinds of aquatic plants are comprised under the name of Algæ, plural of *Alga*, the Latin word for *Sea-weed*. The Algæ take every variety of form. Some have the appearance of a frond, others resemble a flattened leaf, a smooth ribbon, or a round cord. Occasionally they are like globular fibres either striking out singly or disposed as a string of beads. In colour they are generally olive-green, pinkish-purple, crimson, or of a bright grassy-green. Those plants often grow floating in the water without any support, but they are usually fixed to the bottom of the sea, or to rocks under water, by the base of their fronds, which branches out in such a manner as to take a firm hold. If the fronds receive a wound, numbers of young shoots are thrown out from the injured parts, so great is the vitality of this tribe. (Fig. 4.)

In some of the western islands of Scotland, the Bladder-Fucus or Sea-Wrack often serves as winter food for the cattle, which go down to the shore to eat it when the tide is out. Linnæus tells us that in Gothland the people boil this plant in water and mix it with meal to feed their hogs. But one of the most important purposes to which sea-weeds are applied, is the preparation of kelp, a kind of salt which is the principal ingredient in the manufacture of soap, and in making glass.

In Scotland, the Sea-Tangle and the Dulse, two species of Fucus, so called from φῦχος, the Greek word for *Sea-weed*, are dried and eaten. The stems of the Tangle are used for making knife-handles, a thick stem being cut into pieces of the proper length and the hilt of a knife stuck into each. The stem contracts in drying, and hardens firmly round the hilt till it has the appearance of horn. The thicker stems are dried and used for firing.

The size that some of the larger kinds of sea-weeds attain, and the rapidity of their growth, are perfectly astonishing. The Gigantic Fucus is said to extend often to the length of a thousand or fifteen hundred feet, and it grows in such profusion that large masses of it resemble islands. In the Atlantic, Pacific, and Indian Oceans there are vast tracts of sea-weeds, one of which has been called by navigators the Grassy Sea, on account of its great extent. The surface of the sea, in such places, is literally covered with those plants, and ships on their voyages are often several days in steering their way through them.

In China, a species of Fucus is dissolved and employed like glue, as gum is used by us. When washed and steeped in warm water it dissolves, and, as it cools, stiffens into a clear colourless glue, with which large sheets of paper are coated over in order to make them transparent. The paper so prepared is used instead of glass for lanterns and windows. In Japan, windows are sometimes made of slips of split bamboo, which are crossed lozenge-wise, and the spaces between filled up with thin flakes of this glue alone. The nests of a kind of swallow are eaten as a delicacy in China and throughout the East, and are even imported to London. Those nests, it is supposed, are made by the birds almost entirely of a species of sea-weed called the Fucus lichenoïdes, or Lichen-like Fucus. And here it may be observed, that in words which end in *oides*, those five letters should make three syllables, the *i* being long. We ought to say, therefore, li-ken-o-eye-des, and not li-ken-o-ides.

Sometimes, early in Winter or late in the Autumn, we see under trees, and close to park palings, a number of little whitish dots or tubercles peeping up

from amongst the damp grass and decayed leaves.
Each of those tubercles is a young Mushroom, as we
should find if we returned to the spot in the course
of a few days, for plants of the Fungus tribe are
generally of very rapid growth and short duration.
Funguses or Fungi, the *g* soft and the *i* long, have
no leaves, their whole substance being fleshy. Several of the wild funguses are wonderfully beautiful in
colour and form, but some are fearfully poisonous.
The best way to avoid them is to follow the tradi-
tions respecting them of the people of the country
where they grow, and to be guided by their experience. It would, no doubt, be still more prudent
never to venture to eat any sort of Fungus, of the
nature of which we could not feel perfectly sure ourselves. The only kind ever eaten in England is the
field Mushroom, Agaricus campestris. It is cultivated
in hotbeds, and grows wild in parks and fields that
have been for a long time undisturbed by the plough.
Its appearance is by no means striking. The crown,
which is the receptacle, and the foot-stalk, are white,
and what we call the Plates or Gills, which grow
underneath the crown, like a lining, are of a delicate
pink or salmon-colour while young, but become deep
cinnamon-brown as the plant grows older.

At first the gills are concealed by a white membrane composed of parallel threads, none of them crossed, which makes the tissue extremely light and fragile. This membrane extends from the lower edge of the crown to the foot-stalk like a sort of bag or inverted umbrella. As the Mushroom developes it falls off, and nothing remains of it but a sort of ruffle or frill round the foot-stalk, near its top.

When the Mushroom is very young it is enclosed in a white membrane that wraps it up, as if in a bag, from the top of the crown to the bottom of the foot-stalk. This covering is called the Volva, which is the Latin word for a *wrapper* or *covering*. The smaller membrane that connects the crown with the footstalk is called the Cortina ; *cortina* being the Latin for the *cover of a caldron-shaped vessel*.

When the fine threads, from which the young foot-stalks spring, are not placed in a position favourable to the growth of those foot-stalks, they take the name of Byssus. What is called the Dry-rot in wood is said to be occasioned by peculiar kinds of Byssus, of which the texture is so light as to be dispersed by the slightest breath, like the finest cobweb, or wool; and yet, in time, they can destroy the hardest and closest grained wood.

When we look at the surface of one of the gills of the Mushroom through a magnifying glass, we see nothing but that it has a velvety appearance. It is only with the help of a very powerful microscope that its real structure can be discovered. It is then found that it consists of distinct layers of cells, some of those which are tallest, on each surface of the gill, terminating in four points, each point supporting a minute globular bag of a brown colour. Those little brown bags, four and four of them at the extremities of some of the cells, are called spore-cases. The brown colour of the plate or gill in its mature state is owing to their presence. (Fig. 5.) The spores of some of the Fungus family are so exceedingly minute, that they resemble smoke or vapour rather than fine dust. This you can see in Puff-balls and Earth-Stars. The spore-case in the Earth-Star is a brown bag-like cell, in shape like a hyacinth-bulb, with a hole at the top through which the ripe spores escape, as smoke escapes through a chimney.

The orange-coloured or rusty-looking spots that we often perceive on the under surface of the leaves of a Rose tree, are groups of a parasitical Fungus, called Urédo. It is at first developed in the green pulpy substance of the leaf beneath the skin, and at a later period of its growth it takes the form of a flattened tubercle covered with cells of different colours and forms. (Fig. 6.)

The best and most original works that relate to the Fungus family, are "Recherches sur l'Hymenium des Champignons," and "Recherches sur le developpement des Urédinées," by Dr. Leveillé.

What we call Mouldiness is a sort of Fungus of which the spores float in the air till they fix themselves upon different substances and fluids, such as paste, preserved fruit, and ink, where they take root and spread rapidly. Some of the species are composed of a little stalk with a very tiny globular head filled with granules of a green colour, others are tufts of a soft cotton-like substance on a blue and yellow sort of crust, and some consist of little cells ranged in a circle, which burst, and disperse the spores as fast as they ripen.

Mouldiness and Mildew are effectually prevented by any kind of perfume, or essential oil. It is well known that books bound in Russia leather are never touched by Mildew.

In the structure of the several plants of which we have spoken, the order and regularity, the exceeding fitness of the different parts for their appointed work, strike us as something marvellous. The most lavish richness of texture, form, and colour, and yet no jostling, no confusion, no waste of materials or of space. When Galen, one of the oldest Greek philosophers, was giving a public lecture on the structure of the hand and foot, he said: " In explaining these things, I esteem myself as composing a solemn hymn to the Great Maker of our bodily frame, in which I think there is more true piety than in sacrificing hecatombs of oxen, or in burning the most costly perfumes; for I first endeavour from His works to know Him myself, and afterwards, by the same means, to show Him to others, to inform them how great is His wisdom, His goodness, His power." Such were the ideas suggested by the study of anatomy to the mind of a pagan—ideas alike reverent and ennobling; but we, as Christians, can learn a still diviner lesson when we consider the lilies of the field how they grow, and are reminded of Him who told us to emulate their contentment and quiet happiness.

THE END.

LONDON :
R. CLAY, SON. AND TAYLOR, PRINTERS,
BREAD STREET HILL.

ORIGINAL JUVENILE LIBRARY.

A CATALOGUE

OF

NEW AND POPULAR WORKS,

PRINCIPALLY FOR THE YOUNG.

Goldsmith Introduced to Newbery by Dr. Johnson.

PUBLISHED BY

GRIFFITH AND FARRAN,

(SUCCESSORS TO NEWBERY AND HARRIS),

CORNER OF ST. PAUL'S CHURCHYARD,
LONDON.

WERTHEIMER AND CO., CIRCUS PLACE, FINSBURY CIRCUS.

STANESBY'S ILLUMINATED GIFT BOOKS.

Every page richly printed in Gold and Colours.

The Floral Gift.

Small 4to, price 14s. cloth elegant; 21s. morocco extra.

" This is indeed an elegant gift book. Every page has a border printed in Gold and Colours, in which our chief floral favourites are admirably depicted. The text is worthy of the illustrations, the binding is gorgeous, yet in good taste."—*Gentleman's Magazine.*

Aphorisms of the Wise and Good.

With a Photographic Portrait of Milton; intended as a companion volume to "Shakespeare's Household Words." Price 9s. cloth, elegant, 14s. Turkey morocco antique.

" A perfect gem in binding, illustratilon, and literary excellence."—*Daily News.*

Shakespeare's Household Words; :

With a Photographic Portrait taken from the Monument at Stratford-on-Avon. Price 9s. cloth elegant; 14s. morocco antique.

" An exquisite little gem, fit to be the Christmas offering to Titania or Queen Mab."—*The Critic.*

The Wisdom of Solomon;

From the Book of Proverbs. With a Photographic Frontispiece, representing the Queen of Sheba's visit to Solomon. Small 4to, price 14s. cloth elegant; 18s. calf; 21s. morocco antique.

" The borders are of surprising richness and variety, and the colours beautifully blended."—*Morning Post.*

The Bridal Souvenir;

Containing the Choicest Thoughts of the Best Authors, in Prose and Verse. New Edition, with a Portrait of the Princess Royal. Elegantly bound in white and gold, price 21s.

"A splendid specimen of decorative art, and well suited for a bridal gift."—*Literary Gazette.*

The Birth-Day Souvenir;

A Book of Thoughts on Life and Immortality, from Eminent Writers Small 4to. price 12s. 6d. illuminated cloth; 18s. morocco antique.

" The illuminations are admirably designed."—*Gentleman's Magazine.*

Light for the Path of Life;

From the Holy Scriptures. Small 4to, price 12s. cloth elegant; 15s. calf gilt edges; 18s. morocco antique.

"A fit gift from a loving husband, or from aged friend to youthful favourite."—*Illustrated News.*

NEW AND POPULAR WORKS.

A SPLENDID GIFT BOOK.

Dedicated by Permission to H.R.H. The Princess Royal. In Royal 4to., Elegantly bound in cloth, gilt edges. Price Two Guineas.

The Year: its Leaves and Blossoms;

Illustrated by HENRY STILKE, in Thirteen Beautiful Plates, executed in the highest style of Chromo-Lithographic Art, with Verses from the Poets.

"A charming Gift Book, and sure to be heartily welcomed by the refined 'public,' for whom it is intended."—*Art Union.*

NEW WORK BY JOHN TIMBS.

Nooks and Corners of English Life.

Past and Present. By JOHN TIMBS, Author of "Strange Stories of the Animal World." With Illustrations. Post 8vo, price 6s. cloth; 6s. 6d. gilt edges.

NEW WORK BY CHARLES BENNETT.

Lightsome and the Little Golden Lady.

Written and Illustrated by C. H. BENNETT. Twenty-four Engravings. Fcap. 4to., price 3s. 6d. cloth elegant; 4s. 6d. coloured, gilt edges.

NEW WORK BY CAPTAIN MARRYAT'S DAUGHTER.

The Early Start in Life.

By EMILIA MARRYAT NORRIS. With Illustrations by J. LAWSON. Post 8vo, price 5s. cloth elegant; 5s. 6d. gilt edges.

Casimir, the Little Exile.

By CAROLINE PEACHEY. With Illustrations by C. STANTON. Post 8vo., price 4s. 6d. cloth elegant; 5s. gilt edges.

Lucy's Campaign;

A Story of Adventure. By MARY and CATHERINE LEE. With Illustrations by GEORGE HAY. Fcap. 8vo, price 3s. cloth elegant; 3s. 6d. gilt edges.

The Holidays Abroad;

Or, Right at Last. By EMMA DAVENPORT, Author of "Our Birthdays," etc. With Frontispiece by G. HAY. Fcap. 8vo., price 2s. 6d. cloth extra; 3s. gilt edges.

Gerty and May.

By the Author of "Granny's Story Box." With Illustrations by M. L. Vining. Super-royal 16mo, price 2s. 6d. cloth extra; 3s. 6d. coloured, gilt edges.

NEW WORK BY THE HON. MISS BETHELL.

Helen in Switzerland.

By the Hon. Augusta Bethel, Author of "The Echoes of an Old Bell." With Illustrations by E. Whymper. Super-royal 16mo, price 3s. 6d. cloth extra; 4s. 6d. coloured, gilt edges.

WORK BY HOOD'S DAUGHTER.

Wild Roses;

Or, Simple Stories of Country Life. By Francis Freeling Broderip. With Illustrations by H. Anelay. Post 8vo, price 3s. 6d. cloth elegant; 4s. gilt edges.

Nursery Times;

Or, Stories about the Little Ones. By an Old Nurse. With Illustrations by J. Lawson. Imperial 16mo, price 3s. 6d. cloth; 4s. 6d. coloured, gilt edges.

Contents:—1. Early Times. 2. Awkward Times. 3. Happy Times. 4. Troublesome Times. 5. Christmas—Once upon a Time.

The Surprising Adventures of the Clumsy Boy

CRUSOE. By Charles H. Ross. With Twenty-three Coloured Illustrations. Imperial 8vo, price 2s.

Infant Amusements;

Or, How to Make a Nursery Happy. With Hints to Parents and Nurses on the Moral and Physical Training of Children. By W. H. G. Kingston. Post 8vo, price 3s. 6d. cloth.

Taking Tales for Cottage Homes;

in Plain Language and Large Type. To be published in Monthly Numbers, each containing Sixty-four pages, and several Engravings. Crown 8vo, price 4d. each.

No. 1. The MILLER of HILLBROOK; a Tale of English Country Life. (*December* 1.)

No. 2. TOM TRUEMAN. The Life of a Sailor in the Merchant Service. (*January* 1, 1867).

Strange Stories of the Animal World; ˙

A Book of Adventures and Anecdotes, and curious Contributions to Natural History. By JOHN TIMBS, author of "Things Not Generally Known," &c., with Seven Illustrations by ZWECKER, &c. Post 8vo., price 6s., cloth, 6s. 6d., gilt edges.

"Among all the books of the season that will be studied with profit and pleasure, there is not one more meritorious in aim, or more successful in execution."—*Athenæum.*

LADY LUSHINGTON.

Almeria's Castle;

Or, My Early Life in India and England. By LADY LUSHINGTON, Author of "The Happy Home," "Hacco, the Dwarf," &c., with Twelve Illustrations. Super-royal 16mo., price 4s. 6d., cloth, 5s., gilt edges.

"The Authoress has a very graphic pen, and brings before our eyes, with singular vividness, the localities and modes of life she aims to describe."—*London Review.*

Featherland;

Or, How the Birds lived at Greenlawn. By G. W. FENN. With Illustrations by F. W. KEYL. Super-royal 16mo., price 2s. 6d., cloth, 3s. 6d., coloured, gilt edges.

"A delightful book for children. There is no story, but the happiest perception of childish enjoyment is contained in fanciful sketches of bird-life."—*Examiner.*

THOMAS HOOD'S DAUGHTER.

Mamma's Morning Gossips;

Or, Little Bits for Little Birds. Being Easy Lessons for One Month in Words of One Syllable, and a Story to read for each Week. By Mrs. BRODERIP. With Fifty Illustrations by her Brother, THOMAS HOOD. Foolscap Quarto, price 3s., cloth, 4s. 6d. coloured, gilt edges.

"A perfectly delightful reading-book for the little ones."—*Guardian.*

The Australian Babes in the Wood;

A True Story told in Rhyme for the Young. By the Author of "Little Jessie." With Fourteen Engravings from Drawings by HUGH CAMERON, A.R.S.A.; J. McWHIRTIE; GEO. HAY; J. LAWSON, &c. Beautifully printed. 1s. 6d. Boards. 2s. Cloth, gilt edges.

Trottie's Story Book;

True Tales in Short Words and Large Type. By the author of "Tiny Stories," "Tuppy," &c. With Eight Illustrations by WEIR. Super-royal 16mo., price 2s. 6d., cloth, 3s. 6d., coloured, gilt edges.

Six Months in Freshwater;

A Sea-side Tale for Children. With Frontispiece. Super-royal 16mo., price 3s. 6d., cloth.

Work in the Colonies;

Some Account of the Missionary operations of the Church of England in connexion with the Society for the Propagation of the Gospel in Foreign Parts. With Map and Sixteen Illustrations. Royal 16mo., price 5s., cloth.

The Fairy Tales of Science.

A Book for Youth. By J. C. BROUGH. With 16 Beautiful Illustrations by C. H. BENNETT. New Edition, Revised throughout by the author. Fcap. 8vo, price 5s., cloth; 5s. 6d. gilt edges.

"Science, perhaps, was never made more attractive and easy of entrance into the youthful mind."—*The Builder.*

"Altogether the volume is one of the most original, as well as one of the most useful, books of the season."—*Gentleman's Magazine.*

Early Days of English Princes;

By Mrs. RUSSELL GRAY. Illustrations by JOHN FRANKLIN. New and Enlarged Edition. Super-royal 16mo., price 3s. 6d., cloth, 4s. 6d., coloured, gilt edges.

Merry Songs for Little Voices;

The words by Mrs. BRODERIP; set to music by THOMAS MURBY, with 40 illustrations by THOMAS HOOD. Fcap. 4to, price 5s. cloth.

"The merriment is not without meaning or moral, and the songs are enlivened by quaint little cuts."—*Saturday Review.*

THE HONBLE. MISS BETHELL.

Echoes of an Old Bell;

And other Tales of Fairy Lore, by the Honble. AUGUSTA BETHELL. Illustrations by F. W. KEYL. Super royal 16mo., price 3s. 6d. cloth, 4s. 6d. coloured, gilt edges.

"A delightful book of well-conceived and elegantly-written fairy tales."—*Literary Churchman.*

The Primrose Pilgrimage.

A Woodland Story, by M. BETHAM EDWARDS, author of, "Little Bird Blue," "Holidays among the Mountains," etc., with illustrations by T. R. MACQUOID. Imperial 16mo., price 2s. 6d. cloth, 3s. 6d. coloured, gilt edges.

"One of the best books of children's verse that has appeared since the early days of Mary Howitt."—*Nonconformist.*

"The Poems are full of interest, and the Illustrations charming."—*Art Journal.*

Pictures of Girl Life.

By CATHARINE AUGUSTA HOWELL, author of "Pages of Child Life." Frontispiece by F. ELTZE. Fcap. 8vo., price 3s. cloth, 3s. 6d. gilt edges.

"A really healthy and stimulating book for girls."—*Nonconformist.*

The Four Seasons.

A Short Account of the Structure of Plants, being Four Lectures written for the Working Men's Institute, Paris. With Illustrations. Imperial 16mo. Price, 3s 6d. cloth.

"Distinguished by extreme clearness, and teem with information of a useful and popular character."—*Guardian.*

Fun and Earnest;

Or, Rhymes with Reason, by D'ARCY W. THOMPSON, author of "Nursery Nonsense; or, Rhymes without Reason." Illustrated by CHARLES BENNETT. Imperial 16mo., price 3s. cloth, 4s. 6d. coloured. Cloth, Elegant gilt edges.

"Only a clever man with the touch of a poet's feeling in him, can write good children's nonsense; such a man the author proves himself to be."—*Examiner.*

Nursery Nonsense;

Or Rhymes without Reason, by D'ARCY W. THOMPSON, with sixty Illustrations, by C. H. BENNETT. Second edition. Imperial 16mo., price 2s. 6d. cloth; or 4s. 6d. coloured, cloth elegant, gilt edges.

"The funniest book we have seen for an age, and quite as harmless as hearty."—*Daily Review.*
"Whatever Mr. Bennett does, has some touch in it of a true genius."—*Examiner.*

Spectropia;

Or, Surprising Spectral Illusions, showing Ghosts everywhere and of any Colour. By J. H. BROWN. Fourth edition. Quarto. Coloured Plates. Price 2s. 6d. fancy boards.

"One of the best scientific toy books we have seen."—*Athenæum.*
"A clever book. The illusions are founded on true scientific principles."—*Chemical News.*
"We heartily commend Mr. Brown's ingenious work."—*The Lancet.*

BY THE AUTHOR OF 'MARY POWELL,' ETC.
The Interrupted Wedding;

A Hungarian Tale. With Frontispiece, by HENRY WARREN. Post 8vo., price 6s. cloth.

"The author treads on fresh ground, and introduces us to a people of whose home scenes we are glad to read such truthful natural descriptions."—*Athenæum.*
The story is excellently told, as might be expected from the peculiar powers of the narrator."—*Saturday Review.*

BY MRS. HENRY WOOD.
William Allair;

Or, Running away to Sea, by Mrs. H. WOOD, author of "East Lynne," "The Channings," etc. Frontispiece by F. GILBERT. Fcap. 8vo., price 2s. 6d., cloth, 3s. gilt edges.

"There is a fascination about Mrs. Wood's writings, from which neither old nor young can escape."—*Bell's Messenger.*

LADY LUSHINGTON.

Hacco the Dwarf;

Or, The Tower on the Mountain ; and other Tales, by LADY LUSH-
INGTON, author of " The Happy Home." Illustrated by G. J. PINWELL.
Super royal 16mo., price 3s. 6d. cloth, 4s. 6d. coloured, gilt edges.

" Enthusiasm is not our usual fashion, but the excellence of these stories is so greatly
above the average of most clever tales for the play-room, that we are tempted to reward
the author with admiration."--*Athenæum*.

The Happy Home;

Or the Children at the Red House, by LADY LUSHINGTON. Illustrated
by G. J. PINWELL. Super royal 16mo., price 3s. 6d. cloth, 4s. 6d.
coloured, gilt edges.

" A happy mixture of fact and fiction. Altogether it is one of the best books of the
kind we have met with."—*Guardian*.

The Happy Holidays;

Or, Brothers and Sisters at Home, by EMMA DAVENPORT. Frontispiece
by F. GILBERT. Fcap. 8vo., price 2s. 6d. cloth, 3s. gilt edges.

Our Birth Days;

And how to improve them, by Mrs. E. DAVENPORT, author of " Fickle
Flora," etc. Frontispiece by D. H. FRISTON. Fcap. 8vo., price
2s. 6d. cloth, 3s. gilt edges.

" Most admirably suited as a gift to young girls."—*British Mother's Magazine*.

Fickle Flora,

and her Sea Side Friends. By EMMA DAVENPORT, author of " Live
Toys," etc. With Illustrations by J. Absolon. Super Royal 16mo.
price 3s. 6d. cloth; 4s. 6d. coloured, gilt edges.

Live Toys;

Or, Anecdotes of our Four-legged and other Pets. By EMMA DAVEN-
PORT. With Illustrations by HARRISON WEIR. Second Edition.
Super Royal 16mo. price 2s. 6d. cloth; 3s. 6d. coloured, gilt edges.

" One of the best kind of books for youthful reading."—*Guardian*.

Tiny Stories for Tiny Readers in Tiny Words.

By the author of " Tuppy," " Triumphs of Steam," &c., with Twelve
Illustrations, by HARRISON WEIR. Second edition. Super Royal
16mo., price 2s. 6d. cloth, 3s. 6d. coloured, cloth, elegant gilt edges.

DEDICATED BY PERMISSION TO ROSSINI.

Little by Little.

A series of Graduated Lessons in the Art of Reading Music. Second
Edition. Oblong 8vo., price 3s. 6d. cloth.

" One of the best productions of the kind which have yet appeared."—*Charles Steggall*,
Mus. D., Cantab.

Historical Tales of Lancastrian Times.

By the Rev. H. P. DUNSTER, M.A., with illustrations by JOHN FRANKLIN. Fcap. 8vo., price 5s. cloth, 5s. 6d. gilt edges.

"A volume skilfully treated."—*Saturday Review.*

"Conveys a good deal of information about the manners and customs of England and France in the 15th Century."—*Gentlemen's Magazine.*

Memorable Battles in English History.

Where Fought, why Fought, and their Results. With Lives of the Commanders. By W. H. DAVENPORT ADAMS, author of "Neptune's Heroes; or, the Sea-kings of England." Frontispiece by ROBERT DUDLEY. Post 8vo. price 7s. 6d. extra cloth.

"Of the care and honesty of the author's labours, the book gives abundant proof."—*Athenæum.*

The Loves of Tom Tucker and Little Bo-Peep.

Written and Illustrated by THOMAS HOOD. Quarto, price 2s. 6d. coloured plates.

"Full of fun and of good innocent humour. The Illustrations are excellent."—*The Critic.*

Scenes and Stories of the Rhine.

By M. BETHAM EDWARDS, author of "Holidays among the Mountains," etc. With Illustrations by F. W. KEYL. Super Royal 16mo. price 3s. 6d. cloth; 4s. 6d. coloured, gilt edges.

"Full of amusing incidents, good stories, and sprightly pictures."—*The Dial.*

Holidays Among the Mountains;

Or, Scenes and Stories of Wales. By M. BETHAM EDWARDS. Illustrated by F. J. SKILL. Super royal 16mo.; price 3s. 6d. cloth; 4s. 6d. coloured, gilt edges.

Nursery Fun;

Or, the Little Folks' Picture Book. The Illustrations by C. H. BENNETT. Quarto, price 2s. 6d. coloured plates.

"Will be greeted with shouts of laughter in any nursery."—*The Critic.*

Play-Room Stories;

Or, How to make Peace. By GEORGIANA M. CRAIK. With Illustrations by C. GREEN. Super Royal 16mo. price 3s. 6d. cloth; 4s. 6d. coloured, gilt edges.

"This Book will come with 'peace' upon its wings into many a crowded playroom."—*Art Journal.*

ALFRED ELWES' BOOKS FOR BOYS.

With Illustrations, Fcap. 8vo. price 5s. each cloth; 5s. 6d. gilt edges.

Luke Ashleigh;

Or, School Life in Holland. Illustrated by G. DU MAURIER.

"The author's best book, by a writer whose popularity with boys is great."—*Athenæum.*

Guy Rivers;

Or, a Boy's Struggles in the Great World. Illustrations by H. ANELAY.

Ralph Seabrooke;

Or, The Adventures of a Young Artist in Piedmont and Tuscany. Illustrated by DUDLEY.

Frank and Andrea;

Or Forest Life in the Island of Sardinia. Illustrated by DUDLEY.

Paul Blake;

Or, the Story of a Boy's Perils in the Islands of Corsica and Monte Cristo. Illustrated by H. ANELAY.

WILLIAM DALTON'S BOOKS FOR BOYS.

With Illustrations; Fcap. 8vo. price 5s. each cloth; 5s. 6d. gilt edges.

Lost in Ceylon;

The Story of a Boy and Girl's Adventures in the Woods and Wilds of the Lion King of Kandy. Illustrated by WEIR.

"Clever, exciting and full of true descriptions of the creatures and sights in that noble island."—*Literary Gazette.*

The White Elephant;

Or the Hunters of Ava, and the King of the Golden Foot. Illustrated by WEIR.

"Full of dash, nerve and spirit, and withal freshness."—*Literary Gazette.*

The War Tiger;

Or, The Adventures and Wonderful Fortunes of the Young Sea-Chief and his Lad Chow. Illustrated by H. S. MELVILLE.

"A tale of lively adventure vigorously told, and embodying much curious information." *Illustrated News.*

The Faithful Hound.

A Story in Verse, founded on fact. By LADY THOMAS. With Illustrations by H. WEIR. Imperial 16mo, price 2s. 6d. cloth; 3s. 6d. coloured, gilt edges.

Jack Frost and Betty Snow;

With other Tales for Wintry Nights and Rainy Days. Illustrated by H. Weir. Second Edition. 2s. 6d. cloth; 3s. 6d. coloured, gilt edges

"The dedication of these pretty tales, prove by whom they are written; they are indelibly stamped with that natural and graceful method of amusing while instructing, which only persons of genius possess."—*Art Journal.*

THOMAS HOOD'S DAUGHTER.

Crosspatch, the Cricket, and the Counterpane ;

A Patchwork of Story and Song, by FRANCES FREELING BRODERIP. Illustrated by her brother THOMAS HOOD. Super royal 16mo. price 3s. 6d. cloth, 4s. 6d. coloured, gilt edges.

"Hans Andersen has a formidable rival in this gentle Lady."—*Art Journal.*

My Grandmother's Budget

of Stories and Verses. By FRANCES FREELING BRODERIP. Illustrated by her brother, THOMAS HOOD. Price 3s. 6d. cloth; 4s. 6d. coloured, gilt edges.

"Some of the most charming little inventions that ever adorned the department of literature."—*Illustrated Times.*

Tiny Tadpole;

And other Tales. By FRANCES FREELING BRODERIP, daughter of the late Thomas Hood. With Illustrations by HER BROTHER. Super-Royal 16mo. price 3s. 6d. cloth; 4s. 6d. coloured, gilt edges.

"A remarkable book, by the brother and sister of a family in which genius and fun are inherited."—*Saturday Review.*

Funny Fables for Little Folks.

By FRANCES FREELING BRODERIP. Illustrated by her Brother. Super Royal 16mo. price 2s. 6d. cloth; 3s. 6d. coloured, gilt edges.

"The Fables contain the happiest mingling of fun, fancy, humour, and instruction."—*Art Journal.*

CAPTAIN MARRYAT'S DAUGHTER.

What became of Tommy;

By EMILIA MARRYAT NORRIS. With Illustrations by ABSOLON. Super-royal 16mo., price 2s. 6d., cloth, 3s. 6d., coloured, gilt edges.

A Week by Themselves ;

By EMILIA MARRYAT NORRIS, with illustrations by CATHARINE A. EDWARDS. Super royal 16mo., price 2s. 6d. cloth, 3s. 6d. coloured, gilt edges.

" Our younger readers will be charmed with a story of some youthful Crusoes, written by the daughter of Captain Marryat."--*Guardian.*

Harry at School ;

By EMILIA MARRYAT. With Illustrations by ABSOLON. Super Royal 16mo. price 2s. 6d. cloth; 3s. 6d. coloured, gilt edges.

Long Evenings ;

Or, Stories for My Little Friends, by EMILIA MARRYAT. Illustrated by ABSOLON. Second Edition. Price 2s. 6d. cloth; 3s. 6d. coloured, gilt edges.

LANDELL'S INSTRUCTIVE AND AMUSING WORKS.

The Boy's own Toy Maker.

A Practical Illustrated Guide to the useful employment of Leisure Hours. By E. LANDELLS. With Two Hundred Cuts. Sixth Edition. Royal 16mo, price 2s. 6d., cloth.

" A new and valuable form of endless amusement."—*Nonconformist.*
" We recommend it to all who have children to be instructed and amused."—*Economist.*

The Girl's Own Toy Maker,

And Book of Recreation. By E. and A. LANDELLS. Third Edition. With 200 Illustrations. Royal 16mo. price 2s. 6d. cloth.

" A perfect magazine of information."—*Illustrated News of the World.*

Home Pastime ;

Or, The Child's Own Toy Maker. With practical instructions. By E. LANDELLS. New and Cheaper Edition, price 3s. 6d. complete, with the Cards, and Descriptive Letterpress.

. By this novel and ingenious "Pastime," Twelve beautiful Models can be made by Children from the Cards, by attending to the Plain and Simple Instructions in the Book.

" As a delightful exercise of ingenuity, and a most sensible mode of passing a winter's evening, we commend the Child's own Toy Maker."—*Illustrated News.*
" Should be in every house blessed with the presence of children."—*The Field.*

The Illustrated Paper Model Maker ;

Containing Twelve Pictorial Subjects, with Descriptive Letter-press and Diagrams for the construction of the Models. By E. LANDELLS. Price 2s. in a neat Envelope.

" A most excellent mode of educating both eye and hand in the knowledge of form."—*English Churchman.*

THE LATE THOMAS HOOD.

Fairy Land;

Or, Recreation for the Rising Generation, in Prose and Verse. By THOMAS and JANE HOOD. Illustrated by T. HOOD, Jun. Second Edition. Super-royal 16mo; price 3s. 6d. cloth; 4s. 6d. coloured gilt edges.

"These tales are charming. Before it goes into the Nursery, we recommend all grown up people should study ' Fairy Land'—*Blackwood.*'"

The Headlong Career and Woful Ending of Preco-

CIOUS PIGGY. Written for his Children, by the late THOMAS HOOD. With a Preface by his Daughter; and Illustrated by his Son. Fourth Edition. Post 4to, fancy boards, price 2s. 6d., coloured.

"The Illustrations are intensely humourous."—*The Critic.*

BY THE AUTHOR OF " TRIUMPHS OF STEAM," ETC.

Meadow Lea;

Or, the Gipsy Children; a Story founded on fact. By the Author of " The Triumphs of Steam," " Our Eastern Empire," etc. With Illustrations by JOHN GILBERT. Fcap. 8vo. price 4s. 6d. cloth; 5s. gilt edges.

The Triumphs of Steam;

Or, Stories from the Lives of Watt, Arkwright, and Stephenson. With Illustrations by J. GILBERT. Dedicated by permission to Robert Stephenson, Esq., M.P. Second edition. Royal 16mo, price 3s. 6d. cloth; 4s. 6d., coloured, gilt edges.

"A most delicious volume of examples."—*Art Journal.*

Our Eastern Empire;

Or, Stories from the History of British India. Second Edition, with Continuation to the Proclamation of Queen Victoria. With Four Illustrations. Royal 16mo. cloth 3s. 6d.; 4s. 6d. coloured, gilt edges.

"These stories are charming, and convey a general view of the progress of our Empire in the East. The tales are told with admirable clearness."—*Athenæum.*

Might not Right;

Or, Stories of the Discovery and Conquest of America. Illustrated by J. Gilbert. Royal 16mo. price 3s. 6d. cloth; 4s. 6d. coloured, gilt edges.

"With the fortunes of Columbus, Cortes, and Pizarro, for the staple of these stories, the writer has succeeded in producing a very interesting volume."—*Illustrated News.*

Tuppy;

Or the Autobiography of a Donkey. By the Author of " The Triumphs of Steam," etc., etc. Illustrated by HARRISON WEIR. Super Royal 16mo. price 2s. 6d. cloth; 3s. 6d. coloured, gilt edges.

"A very intelligent donkey, worthy of the distinction conferred upon him by the artist."—*Art Journal.*

1. The History of a Quartern Loaf.

in Rhymes and Pictures. By WILLIAM NEWMAN. 12 Illustrations.
Price 6d. plain, 1s. coloured. 2s. 6d. on linen, and bound in cloth.

Uniform in size and price,

2. The History of a Cup of Tea.

3. The History of a Scuttle of Coals.

4. The History of a Lump of Sugar.

5. The History of a Bale of Cotton.

6. The History of a Golden Sovereign.

*** Nos. 1 to 3 and 4 to 6, may be had bound in Two Volumes. Cloth
price 2s. each, plain; 3s. 6d. coloured.

Distant Homes;

Or, the Graham Family in New Zealand. By Mrs. I. E. AYLMER.
With Illustrations by J. JACKSON. Super Royal 16mo. price 3s. 6d.
cloth; 4s. 6d. coloured, gilt edges.

"English children will be delighted with the history of the Graham Family, and be
enabled to form pleasant and truthful conceptions of the 'Distant Homes' inhabited by
their kindred."—*Athenæum.*

Neptune's Heroes : or The Sea Kings of England;

from Hawkins to Franklin. By W. H. DAVENPORT ADAMS. Illustrated
by MORGAN. Fcap. 8vo; price 5s. cloth; 5s. 6d. gilt edges.

"We trust Old England may ever have writers as ready and able to interpret to her
children the noble lives of her greatest men."—*Athenæum.*

Hand Shadows,

To be thrown upon the Wall. By HENRY BURSILL. First and Second
Series each containing Eighteen Original Designs. 4to price 2s. each
plain; 2s. 6d. coloured.

"Uncommonly clever—some wonderful effects are produced."—*The Press.*

BY W. H. G. KINGSTON.

Our Soldiers;

Or, Anecdotes of the Campaigns and Gallant Deeds of the British Army during the reign of Her Majesty Queen Victoria. By W. H. G. KINGSTON. With Frontispiece from a Painting in the Victoria Cross Gallery. Second Edition. Fcp. 8vo. price 3s. cloth; 3s. 6d. gilt edges.

Our Sailors;

Or, Anecdotes of the Engagements and Gallant Deeds of the British Navy during the reign of Her Majesty Queen Victoria. By W. H. G. KINGSTON. With Frontispiece. Second Edition. Fcap. 8vo. price 3s. cloth; 3s. 6d. gilt edges.

"These volumes abundantly prove that both our officers and men in the Army and Navy, have been found as ready as ever to dare, and to do as was dared and done of yore, when led by a Nelson or a Wellington."

W. H. G. KINGSTON'S BOOKS FOR BOYS.

With Illustrations. Fcap. 8vo. price 5s. each, cloth; 5s. 6d. gilt edges.

True Blue;

Or, the Life and Adventures of a British Seaman of the Old School.

"There is about all Mr. Kingston's tales a spirit of hopefulness, honesty, and cheery good principle, which makes them most wholesome, as well as most interesting reading."—*Era.*

"With the exception of Capt. Marryat, we know of no English author who will compare with Mr. Kingston as a writer of books of nautical adventure."—*Illustrated News.*

Will Weatherhelm;

Or, the Yarn of an Old Sailor about his Early Life and Adventures.

Fred Markham in Russia;

Or, the Boy Travellers in the Land of the Czar.

Salt Water;

Or Neil D'Arcy's Sea Life and Adventures. With Eight Illustrations.

Mark Seaworth;

A Tale of the Indian Ocean. With Illustrations by J. ABSOLON. Second Edition.

Peter the Whaler;

His early Life and Adventures in the Arctic Regions. Third Edition. Illustrations by E. DUNCAN.

Old Nurse's Book of Rhymes, Jingles, and Ditties.

Illustrated by C. II. BENNETT. With Ninety Engravings. New Edition. Fcap. 4to., price 3s. 6d. cloth, plain, or 6s. coloured.

"The illustrations are all so replete with fun and imagination, that we scarcely know who will be most pleased with the book, the good-natured grandfather who gives it, or the chubby grandchild who gets it, for a Christmas-Box."—*Notes and Queries.*

Home Amusements.

A Choice Collection of Riddles, Charades, Conundrums, Parlour Games, and Forfeits. By PETER PUZZLEWELL, Esq., of Rebus Hall. New Edition, with Frontispiece by PHIZ. 16mo, 2s. 6d. cloth.

Clara Hope;

Or, the Blade and the Ear. By MISS MILNER. With Frontispiece by Birket Foster. Fcap. 8vo. price 3s. 6d. cloth; 4s. 6d. cloth elegant, gilt edges.

"A beautiful narrative, showing how bad habits may be eradicated, and evil tempers subdued."—*British Mother's Journal.*

Pages of Child Life;

By CATHARINE AUGUSTA HOWELL, author of "Pictures of School Life." With Three Illustrations. Fcap. 8vo., price 3s. 6d. cloth.

The Adventures and Experiences of Biddy Dork-

ING and of the FAT FROG. Edited by MRS. S. C. HALL. Illustrated by H. Weir. 2s. 6d. cloth; 3s. 6d. coloured, gilt edges.

"Most amusingly and wittily told."—*Morning Herald.*

Historical Acting Charades;

Or, Amusements for Winter Evenings, by the author of "Cat and Dog," etc. New Edition. Fcap. 8vo., price 3s. 6d. cloth gilt edges.

"A rare book for Christmas parties, and of practical value."—*Illustrated News.*

The Story of Jack and the Giants:

With thirty-five Illustrations by RICHARD DOYLE. Beautifully printed. New and Cheaper Edition. Fcap. 4to. price 2s. 6d. cloth; 3s. 6d. coloured, extra cloth, gilt edges.

"In Doyle's drawings we have wonderful conceptions, which will secure the book a place amongst the treasures of collectors, as well as excite the imaginations of children."—*Illustrated Times.*

Granny's Wonderful Chair;

And its Tales of Fairy Times. By FRANCES BROWNE. Illustrations by KENNY MEADOWS. 3s. 6d. cloth, 4s. 6d. coloured.

"One of the happiest blendings of marvel and moral we have ever seen."—*Literary Gazette.*

The Early Dawn;

Or, Stories to Think about. Illustrated by II. WEIR, etc. Small 4to.; price 2s. 6d. cloth; 3s. 6d. coloured, gilt edges.

Angelo;

Or, the Pine Forest among the Alps. By GERALDINE E. JEWSBURY, author of "The Adopted Child," etc. Illustrations by J. ABSOLON. Second Edition. Price 2s. 6d. cloth; 3s. 6d. coloured, gilt edges.

"As pretty a child's story as one might look for on a winter's day."—*Examiner.*

Tales of Magic and Meaning.

Written and Illustrated by ALFRED CROWQUILL. Small 4to.; price 3s. 6d. cloth; 4s. 6d. coloured.

"Cleverly written, abounding in frolic and pathos, and inculcates so pure a moral, that we must pronounce him a very fortunate little fellow, who catches these ' Tales of Magic,' as a windfall from ' The Christmas Tree'."—*Athenæum.*

Faggots for the Fire Side;

Or, Tales of Fact and Fancy. By PETER PARLEY. With Twelve Tinted Illustrations. New Edition. Foolscap 8vo.; 3s. 6d., cloth; 4s. 6d. coloured, gilt edges.

"A new book by Peter Parley is a pleasant greeting for all boys and girls, wherever the English language is spoken and read. He has a happy method of conveying information, while seeming to address himself to the imagination."—*The Critic.*

Letters from Sarawak,

Addressed to a Child; embracing an Account of the Manners, Customs, and Religion of the Inhabitants of Borneo, with Incidents of Missionary Life among the Natives. By Mrs. M'DOUGALL. Fourth Thousand, with Illustrations. 3s. 6d. cloth.

" All is new, interesting, and admirably told."—*Church and State Gazette.*

Kate and Rosalind;

Or, Early Experiences. By the author of "Quicksands on Foreign Shores," etc. Fcap. 8vo, 3s. 6d. cloth; 4s. gilt edges.

" A book of unusual merit. The story is exceedingly well told, and the characters are drawn with a freedom and boldness seldom met with."—*Church of England Quarterly.*

" We have not room to exemplify the skill with which Puseyism is tracked and detected. The Irish scenes are of an excellence that has not been surpassed since the best days of Miss Edgeworth."—*Fraser's Magazine.*

Clarissa Donnelly;

Or, The History of an Adopted Child. By Geraldine E. Jewsbury. With an Illustration by John Absolon. Fcap. 8vo, 3s. 6d. cloth; 4s. gilt edges.

"With wonderful power, only to be matched by as admirable a simplicity, Miss Jewsbury has narrated the history of a child. For nobility of purpose, for simple, nervous writing, and for artistic construction, it is one of the most valuable works of the day."—*Lady's Companion.*

The Discontented Children;

And How they were Cured. By M. and E. Kirby. Illustrated by H. K. Browne (Phiz.). Third edition, price 2s. 6d. cloth; 3s. 6d. coloured, gilt edges.

"We know no better method of banishing 'discontent' from school-room and nursery than by introducing this wise and clever story to their inmates."—*Art Journal.*

The Talking Bird;

Or, the Little Girl who knew what was going to happen. By M. and E. Kirby. With Illustrations by H. K. Browne. Second Edition. Price 2s. 6d. cloth; 3s. 6d. coloured, gilt edges.

Julia Maitland;

Or, Pride goes before a Fall. By M. and E. Kirby. Illustrated by Absolon. Price 2s. 6d. cloth; 3s. 6d. coloured, gilt edges.

"It is nearly such a story as Miss Edgeworth might have written on the same theme."—*The Press.*

COMICAL PICTURE BOOKS.

Each with Sixteen large Coloured Plates, price 2s. 6d., in fancy boards, or mounted on cloth, 1s. extra.

Picture Fables.

Written and Illustrated by Alfred Crowquill.

The Careless Chicken;

By the Baron Krakemsides. By Alfred Crowquill.

Funny Leaves for the Younger Branches.

By the Baron Krakemsides, of Burstenoudelafen Castle. Illustrated by Alfred Crowquill.

Laugh and Grow Wise;

By the Senior Owl of Ivy Hall. With Sixteen large coloured Plates. Price 2s. 6d. fancy boards; or 3s. 6d. mounted on cloth.

The Remarkable History of the House that Jack

Built. Splendidly Illustrated and magnificently Illuminated by THE SON OF A GENIUS. Price 2s. in fancy cover.

"Magnificent in suggestion, and most comical in expression!"—*Athenæum*.

A Peep at the Pixies;

Or, Legends of the West. By Mrs. BRAY. Author of "Life of Stothard," "Trelawny," etc., etc. With Illustrations by Phiz. Super-royal 16mo, price 3s. 6d. cloth; 4s. 6d. coloured, gilt edges.

"A peep at the actual Pixies of Devonshire, faithfully described by Mrs. Bray, is a treat. Her knowledge of the locality, her affection for her subject, her exquisite feeling for nature, and her real delight in fairy lore, have given a freshness to the little volume we did not expect. The notes at the end contain matter of interest for all who feel a desire to know the origin of such tales and legends."—*Art Journal*.

A BOOK FOR EVERY CHILD.

The Favourite Picture Book;

A Gallery of Delights, designed for the Amusement and Instruction of the Young. With several Hundred Illustrations from Drawings by J. ABSOLON, H. K. BROWNE (Phiz), J. GILBERT, T. LANDSEER, J. LEECH, J. S. PROUT, H. WEIR, etc. New Edition. Royal 4to., bound in a new and Elegant Cover, price 3s. 6d. plain; 7s. 6d. coloured; 10s. 6d. mounted on cloth and coloured.

Ocean and her Rulers;

A Narrative of the Nations who have held dominion over the Sea; and comprising a brief History of Navigation. By ALFRED ELWES. With Frontispiece. Fcap. 8vo, 5s. cloth; 5s. 6d. gilt edges.

"The volume is replete with valuable and interesting information; and we cordially recommend it as a useful auxiliary in the school-room, and entertaining companion in the library."—*Morning Post*.

Sunday Evenings with Sophia;

Or, Little Talks on Great Subjects. A Book for Girls. By LEONORA G. BELL. Frontispiece by J. ABSOLON. Fcap. 8vo, price 2s. 6d. cloth.

Blind Man's Holiday;

Or Short Tales for the Nursery. By the Author of "Mia and Charlie," "Sidney Grey," etc. Illustrated by John Absolon. Super Royal 16mo. price 3s. 6d. cloth; 4s. 6d. coloured, gilt edges.

The Vicar of Wakefield;

A Tale. By OLIVER GOLDSMITH. Printed by Whittingham. With Eight Illustrations by J. ABSOLON. Square fcap. 8vo, price 5s., cloth; 7s. half-bound morocco, Roxburghe style; 10s. 6d. antique morocco.

Mr. Absolon's graphic sketches add greatly to the interest of the volume: altogether, it is as pretty an edition of the ' Vicar' as we have seen. Mrs. Primrose herself would consider it ' well dressed.'"—*Art Journal.*

" A delightful edition of one of the most delightful of works: the fine old type and thick paper make this volume attractive to any lover of books."—*Edinburgh Guardian.*

The Wonders of Home, in Eleven Stories.

By GRANDFATHER GREY. With Illustrations. Third and Cheaper Edition. Royal 16mo., 2s. 6d. cloth; 3s. 6d. coloured, gilt edges.

" The idea is excellent, and its execution equally commendable. The subjects are well selected, and are very happily told in a light yet sensible manner."—*Weekly News.*

Cat and Dog;

Or, Memoirs of Puss and the Captain. Illustrated by WEIR. Eighth Edition. Super-royal 16mo, 2s. 6d. cloth; 3s. 6d. coloured, gilt edges.

" The author of this amusing little tale is, evidently, a keen observer of nature. The illustrations are well executed; and the moral, which points the tale, is conveyed in the most attractive form."—*Britannia.*

The Doll and Her Friends;

Or, Memoirs of the Lady Seraphina. By the Author of " Cat and Dog." Third Edition. With Four Illustrations by H. K. BROWNE (Phiz). 2s. 6d., cloth; 3s. 6d. coloured, gilt edges.

Tales from Catland;

Dedicated to the Young Kittens of England. By an OLD TABBY. Illustrated by H. WEIR. Fourth Edition. Small 4to, 2s. 6d. plain; 3s. 6d. coloured, gilt edges.

"The combination of quiet humour and sound sense has made this one of the pleasantest little books of the season."—*Lady's Newspaper.*

Scenes of Animal Life and Character.

From Nature and Recollection. In Twenty Plates. By J. B. 4to, price 2s., plain; 2s. 6d., coloured, fancy boards.

" Truer, heartier, more playful, or more enjoyable sketches of animal life could scarcely be found anywhere."—*Spectator.*

Anecdotes of the Habits and Instincts of Animals.
Third Edition. With Illustrations by HARRISON WEIR. Fcap. 8vo, 3s. 6d. cloth; 4s. gilt edges.

Anecdotes of the Habits and Instincts of Birds,
REPTILES, and FISHES. With Illustrations by HARRISON WEIR. Second Edition. Fcap. 8vo, 3s. 6d. cloth; 4s. gilt edges.

"Amusing, instructive, and ably written."—*Literary Gazette.*
"Mrs. Lee's authorities—to name only one, Professor Owen—are, for the most part first-rate.'—*Athenæum.*

Twelve Stories of the Sayings and Doings of
ANIMALS. With Illustrations by J. W. ARCHER. Third Edition. Super-royal 16mo, 2s. 6d. cloth; 3s. 6d. coloured, gilt edges.

Familiar Natural History.
With Forty-two Illustrations from Original Drawings by HARRISON WEIR. Super-royal 16mo, 3s. 6d. cloth; 5s. coloured gilt edges.

* May be had in Two Volumes, 2s. each plain; 2s. 6d. Coloured, Entitled "British Animals and Birds." "Foreign Animals and Birds."

Playing at Settlers;
Or, the Faggot House. Illustrated by GILBERT. Second Edition. Price 2s. 6d. cloth; 3s. 6d. coloured, gilt edges.

Adventures in Australia;
Or, the Wanderings of Captain Spencer in the Bush and the Wilds. Second Edition. Illustrated by PROUT. Fcap. 8vo., 5s. cloth; 5s. 6d. gilt edges.

The African Wanderers;
Or, the Adventures of Carlos and Antonio; embracing interesting Descriptions of the Manners and Customs of the Western Tribes, and the Natural Productions of the Country. Fourth Edition. With Eight Engravings. Fcap. 8vo, 3s. 6d. cloth; 4s. gilt edges.

"For fascinating adventure, and rapid succession of incident, the volume is equal to any relation of travel we ever read."—*Britannia.*

ELEGANT GIFT FOR A LADY.
Trees, Plants, and Flowers;
Their Beauties, Uses and Influences. By Mrs. R. LEE. With beautiful coloured Illustrations by J. ANDREWS. 8vo, price 10s. 6d., cloth elegant, gilt edges.

"The volume is at once useful as a botanical work, and exquisite as the ornament of a boudoir table."—*Britannia.* "As full of interest as of beauty."—*Art Journal.*

Fanny and her Mamma;

Or, Easy Lessons for Children. In which it is attempted to bring Scriptural Principles into daily practice. Illustrated by J. GILBERT. Third Edition. 16mo, 2s. 6d. cloth; 3s. 6d. coloured, gilt edges.

"A little book in beautiful large clear type, to suit the capacity of infant readers, which we can with pleasure recommend."—*Christian Ladies' Magazine.*

Short and Simple Prayers,

For the Use of Young Children. With Hymns. Fifth Edition. Square 16mo, 1s. cloth.

" Well adapted to the capacities of children—beginning with the simplest forms which the youngest child may lisp at its mother's knee, and proceeding with those suited to its gradually advancing age. Special prayers, designed for particular circumstances and occasions, are added. We cordially recommend the book."—*Christian Guardian.*

Mamma's Bible Stories,

For her Little Boys and Girls, adapted to the capacities of very young Children. Twelfth Edition, with Twelve Engravings. 2s. 6d. cloth; 3s. 6d. coloured, gilt edges.

A Sequel to Mamma's Bible Stories.

Fifth Edition. Twelve Illustrations. 2s. 6d. cloth, 3s. 6d. coloured.

Scripture Histories for Little Children.

With Sixteen Illustrations, by JOHN GILBERT. Super-royal 16mo. price 2s. 6d. cloth; 3s. 6d. coloured, gilt edges.

CONTENTS.—The History of Joseph—History of Moses—History of our Saviour—The Miracles of Christ.

Sold separately: 6d. each, plain; 1s. coloured.

The Family Bible Newly Opened;

With Uncle Goodwin's account of it. By JEFFERYS TAYLOR. Frontispiece by J. GILBERT. Fcap. 8vo, 3s. 6d. cloth.

" A very good account of the Sacred Writings, adapted to the tastes, feelings, and intelligence of young people."—*Educational Times.*

Good in Everything;

Or, The Early History of Gilbert Harland. By MRS. BARWELL. Author of "Little Lessons for Little Learners," etc. Second Edition. Illustrations by GILBERT. 2s. 6d. cloth; 3s. 6d., coloured, gilt edges.

" The moral of this exquisite little tale will do more good than a thousand set tasks abounding with dry and uninteresting truisms."—*Bell's Messenger.*

THE FAVOURITE LIBRARY.

A Series of Works for the Young; each Volume with an Illustration by a well-known Artist. Price 1s. cloth.

1. THE ESKDALE HERD BOY. By Lady Stoddart.
2. MRS. LEICESTER'S SCHOOL. By Charles and Mary Lamb.
3. THE HISTORY OF THE ROBINS. By Mrs. Trimmer.
4. MEMOIR OF BOB, THE SPOTTED TERRIER.
5. KEEPER'S TRAVELS IN SEARCH OF HIS MASTER.
6. THE SCOTTISH ORPHANS. By Lady Stoddart.
7. NEVER WRONG; or, THE YOUNG DISPUTANT; and "IT WAS ONLY IN FUN."
8. THE LIFE AND PERAMBULATIONS OF A MOUSE.
9. EASY INTRODUCTION TO THE KNOWLEDGE OF NATURE. By Mrs. Trimmer.
10. RIGHT AND WRONG. By the Author of "Always Happy."
11. HARRY'S HOLIDAY. By Jefferys Taylor.
12. SHORT POEMS AND HYMNS FOR CHILDREN.

The above may be had Two Volumes bound in One, at Two Shillings cloth.

Glimpses of Nature;

And Objects of Interest described during a Visit to the Isle of Wight. Designed to assist and encourage Young Persons in forming habits of observation. By Mrs. Loudon. Second Edition, enlarged. With Forty-one Illustrations. 3s. 6d. cloth.

"We could not recommend a more valuable little volume. It is full of information, conveyed in the most agreeable manner."—*Literary Gazette.*

Tales of School Life.

By Agnes Loudon. With Illustrations by John Absolon. Second Edition. Royal 16mo, 2s. 6d. plain; 3s. 6d. coloured, gilt edges.

"These reminiscences of school days will be recognised as truthful pictures of every-day occurrence. The style is colloquial and pleasant, and therefore well suited to those for whose perusal it is intended."—*Athenæum.*

Kit Bam, the British Sinbad;

Or, the Yarns of an Old Mariner. By MARY COWDEN CLARKE, illustrated by GEORGE CRUIKSHANK. Fcap. 8vo, price 3s. 6d. cloth; 4s. gilt edges.

The Day of a Baby Boy;

A Story for a Young Child. By E. BERGER. With Illustrations by JOHN ABSOLON. Third Edition. Super-royal 16mo, price 2s. 6d. cloth; 3s. 6d. coloured, gilt edges.

"A sweet little book for the nursery."—*Christian Times.*

Harry Hawkins's H-Book;

Shewing how he learned to aspirate his H's. Frontispiece by H. WEIR. Second Edition. Super-royal 16mo, price 6d.

"No family or school-room within, or indeed beyond, the sound of Bow bells, should be without this merry manual."—*Art Journal.*

The Ladies' Album of Fancy Work.

Consisting of Novel, Elegant, and Useful Patterns in Knitting, Netting, Crochet, and Embroidery, printed in Colours. Bound in a beautiful cover. Post 4to, 3s. 6d., gilt edges.

Visits to Beechwood Farm;

Or, Country Pleasures. By CATHERINE M. A. COUPER. Illustrations by ABSOLON. Small 4to, 3s. 6d., plain; 4s. 6d. coloured; gilt edges.

The Modern British Plutarch;

Or, Lives of Men distinguished in the recent History of our Country for their Talents, Virtues and Achievements. By W. C. TAYLOR, LL.D. Author of "A Manual of Ancient and Modern History," etc. 12mo, Second Thousand, with a new Frontispiece. 4s. 6d. cloth; 5s. gilt edges.

"A work which will be welcomed in any circle of intelligent young persons."—*British Quarterly Review.*

Stories of Julian and his Playfellows.

Written by HIS MAMMA. With Four Illustrations by JOHN ABSOLON. Second Edition. Small 4to., 2s. 6d., plain; 3s. 6d., coloured, gilt edges.

The Nine Lives of a Cat;

A Tale of Wonder. Written and Illustrated by C. H. BENNETT. Twenty-four Engravings. price 2s. cloth; 2s. 6d. coloured.

"Rich in the quaint humour and fancy that a man of genius knows how to spare for the enlivenment of children."—*Examiner.*

The Celestial Empire;

or, Points and Pickings of Information about China and the Chinese. By the late "OLD HUMPHREY." With Twenty Engravings from Drawings by W. H. PRIOR. Fcap. 8vo, 3s. 6d., cloth; 4s. gilt edges.

"The book is exactly what the author proposed it should be, full of good information good feeling, and good temper."—*Allen's Indian Mail.*

Maud Summers the Sightless:

A Narrative for the Young. Illustrated by ABSOLON. 3s. 6d. cloth; 4s. 6d. coloured, gilt edges.

London Cries and Public Edifices

Illustrated in Twenty-four Engravings by LUKE LIMNER; with descriptive Letter-press. Square 12mo, 2s. 6d. plain; 5s. coloured.

The Silver Swan;

A Fairy Tale. By MADAME DE CHATELAIN. Illustrated by JOHN LEECH. Small 4to, 2s. 6d. cloth; 3s. 6d. coloured, gilt edges.

A Word to the Wise;

Or, Hints on the Current Improprieties of Expression in Writing and Speaking. By PARRY GWYNNE. 11th Thousand. 18mo. price 6d. sewed, or 1s. cloth. gilt edges.

"All who wish to mind their *p's* and *q's* should consult this little volume."—*Gentleman's Magazine.*

Tales from the Court of Oberon.

Containing the favourite Histories of Tom Thumb, Graciosa and Percinet, Valentine and Orson, and Children in the Wood. With Sixteen Illustrations by CROWQUILL. 2s. 6d. plain; 3s. 6d. coloured.

Rhymes of Royalty.

The History of England in Verse, from the Norman Conquest to the reign of QUEEN VICTORIA; with an Appendix, comprising a summary of the leading events in each reign. Fcap. 8vo, 2s. 6d. cloth.

True Stories from Ancient History,

Chronologically arranged from the Creation of the World to the Death of Charlemagne. Twelfth Edition. With 24 Steel Engravings. 12mo, 5s. cloth.

True Stories from Modern History,

From the Death of Charlemagne to the present Time. Eighth Edition. With 24 Steel Engravings. 12mo, 5s. cloth.

Mrs. Trimmer's Concise History of England,

Revised and brought down to the present time by Mrs. MILNER. With Portraits of the Sovereigns in their proper costume, and Frontispiece by HARVEY. New Edition in One Volume. 5s. cloth.

Stories from the Old and New Testaments,

On an improved plan. By the Rev. B. H. DRAPER. With 48 Engravings. Fifth Edition. 12mo, 5s. cloth.

Wars of the Jews,

As related by JOSEPHUS; adapted to the Capacities of Young Persons, With 24 Engravings. Sixth Edition. 4s. 6d. cloth.

Pictorial Geography.

For the use of Children. Presenting at one view Illustrations of the various Geographical Terms, and thus imparting clear and definite ideas of their meaning. On a Large Sheet. Price 2s. 6d. in tints; 5s. on Rollers, varnished.

One Thousand Arithmetical Tests;

Or, The Examiner's Assistant. Specially adapted for Examination Purposes, but also suited for general use in Schools. By T. S. CAYZER, Head Master of Queen Elizabeth's Hospital, Bristol. Third Edition, revised and stereotyped. Price 1s. 6d. cloth.

₊ Answers to the above, 1s. 6d. cloth.

One Thousand Algebraical Tests;

On the same plan. 8vo., price 3s. 6d. cloth.
Answers to the Algebraical Tests, price 2s. 6d. cloth.

Gaultier's Familiar Geography.

With a concise Treatise on the Artificial Sphere, and two coloured Maps, illustrative of the principal Geographical Terms. Sixteenth Edition. 16mo, 3s. cloth.

Gaultier's Atlas.

Consisting of 8 Maps coloured, and 7 in Outline, etc. Folio, 15s. half-bound.

Butler's Outline Maps, and Key;

Or, Geographical and Biographical Exercises; with a Set of Coloured Outline Maps; designed for the Use of Young Persons. By the late WILLIAM BUTLER. Enlarged by the author's son, J. O. BUTLER. Thirty-third Edition, revised. **4s.**

Every-Day Things;

Or, Useful Knowledge respecting the principal Animal, Vegetable, and Mineral Substances in common use. Second Edition. 18mo, 1s. 6d. cloth.

"A little encyclopædia of useful knowledge, deserving a place in every juvenile library."
—*Evangelical Magazine.*

MARIN DE LA VOYE'S ELEMENTARY FRENCH WORKS.

Les Jeunes Narrateurs;

Ou Petits Contes Moraux. With a Key to the difficult words and phrases. Frontispiece. Second Edition. 18mo, 2s. cloth.
"Written in pure and easy French."—*Morning Post.*

The Pictorial French Grammar;

For the Use of Children. With Eighty Illustrations. Royal 16mo., price 1s. sewed; 1s. 6d. cloth.

Rowbotham's New and Easy Method of Learning
the FRENCH GENDERS. New Edition. **6d.**

Bellenger's French Word and Phrase-book.

Containing a select Vocabulary and Dialogues, for the Use of Beginners. New Edition, 1s. sewed.

Le Babillard.

An Amusing Introduction to the French Language. By a French Lady. Seventh Edition. With 16 Illustrations. 2s. cloth.

Der Schwätzer;

Or, the Prattler. An amusing Introduction to the German Language, on the Plan of "Le Babillard." 16 Illustrations. 16mo, price 2s. cloth.

Battle Fields.

A graphic Guide to the Places described in the History of England as the scenes of such Events; with the situation of the principal Naval Engagements fought on the Coast of the British Empire. By Mr. WAUTHIER, Geographer. On a large sheet 3s. 6d.; in case 6s.; or on a roller, and varnished, 7s. 6d.

Tabular Views of the Geography and Sacred History of PALESTINE, and of the TRAVELS of ST. PAUL.

Intended for Pupil Teachers, and others engaged in Class Teaching. By A. T. WHITE. Oblong 8vo, price 1s., sewed.

The First Book of Geography;

Specially adapted as a Text Book for Beginners, and as a Guide to the Young Teacher. By HUGO REID, author of "Elements of Astronomy," etc. Fourth Edition, carefully revised. 18mo, 1s. sewed.

"One of the most sensible little books on the subject of Geography we have met with." —*Educational Times.*

The Child's Grammar,

By the late LADY FENN, under the assumed name of Mrs. Lovechild. Fiftieth Edition. 18mo, 9d. cloth.

The Prince of Wales' Primer.

With 300 Illustrations by J. GILBERT. New Edition, price 6d.

Always Happy;

Or, Anecdotes of Felix and his Sister Serena. Nineteenth Edition, with Illustrations by ANELAY. Royal 18mo, price 2s. cloth.

Anecdotes of Kings,

Selected from History; or, Gertrude's Stories for Children. With Engravings. 2s. 6d. plain; 3s. 6d. coloured.

Bible Illustrations;

Or, a Description of Manners and Customs peculiar to the East, and especially Explanatory of the Holy Scriptures. By the Rev. B. H. DRAPER. With Engravings. Fourth Edition. Revised by J. KITTO, Editor of "The Pictorial Bible," etc. 3s. 6d. cloth.

The British History briefly told,

and a Description of the Ancient Customs, Sports, and Pastimes of the English. Embellished with Portraits of the Sovereigns of England in their proper Costumes, and 18 other Engravings. 3s. 6d. cloth.

Chit-chat;

Or, Short Tales in Short Words. By the author of " Always Happy." New Edition. With Eight Engravings. Price 2s. 6d. cloth, 3s. 6d. coloured, gilt edges.

Conversations on the Life of Jesus Christ.

By a MOTHER. With 12 Engravings. 2s. 6d. plain; 3s. 6d. coloured.

Cosmorama.

The Manners, Customs, and Costumes of all Nations of the World described. By J. ASPIN. With numerous Illustrations. 3s. 6d. plain; and 4s. 6d. coloured.

Easy Lessons;

Or, Leading-Strings to Knowledge. New Edition, with 8 Engravings. 2s. 6d. plain; 2s. 6d. coloured, gilt edges.

Key to Knowledge;

Or, Things in Common Use simply and shortly explained. By a MOTHER, Author of " Always Happy," etc. Thirteenth Edition. With Sixty Illustrations. 2s. 6d. cloth.

Facts to correct Fancies;

Or, Short Narratives compiled from the Biography of Remarkable Women. By a MOTHER. With Engravings, 3s. 6d. plain; 4s. 6d. coloured.

Fruits of Enterprise;

Exhibited in the Travels of Belzoni in Egypt and Nubia. Fourteenth Edition, with six Engravings by BIRKET FOSTER. Price 3s. cloth.

The Garden;

Or, Frederick's Monthly Instructions for the Management and Formation of a Flower Garden. Fourth Edition. With Engravings by SOWERBY. 3s. 6d. plain; or 6s. with the Flowers coloured.

How to be Happy;

Or, Fairy Gifts: to which is added a Selection of Moral Allegories. With Steel Engravings. Price 3s. 6d. cloth.

Infantine Knowledge.

A Spelling and Reading Book, on a Popular Plan. With numerous Engravings. Tenth Edition. 2s. 6d. plain; 3s. 6d. coloured, gilt edges.

The Ladder to Learning.

A Collection of Fables, arranged progressively in words of One, Two, and Three Syllables. Edited by Mrs. TRIMMER. With 79 Cuts. Nineteenth Edition. 2s. 6d. cloth.

Little Lessons for Little Learners.

In Words of One Syllable. By Mrs. BARWELL. Tenth Edition, with numerous Illustrations. 2s. 6d. plain; 3s. 6d. coloured, gilt edges.

The Little Reader.

A Progressive Step to Knowledge. Fourth Edition with sixteen Plates. Price 2s. 6d. cloth.

Mamma's Lessons.

For her Little Boys and Girls. Thirteenth Edition, with eight Engravings. Price 2s. 6d. cloth; 3s. 6d. coloured, gilt edges.

The Mine;

Or, Subterranean Wonders. An Account of the Operations of the Miner and the Products of his Labours. By the late Rev. ISAAC TAYLOR. Sixth Edition, with numerous additions by Mrs. LOUDON. 45 Woodcuts and 16 Steel Engravings. 3s. 6d. cloth.

Rhoda;

Or, The Excellence of Charity. Fourth Edition. With Illustrations. 16mo, 2s. cloth.

The Rival Crusoes,

And other Tales. By AGNES STRICKLAND, author of "The Queens of England." Sixth Edition. 18mo, price 2s. cloth.

Short Tales.

Written for Children. By DAME TRUELOVE. 20 Engravings. 3s. 6d. cloth.

The Students;

Or, Biographies of the Grecian Philosophers. 12mo, price 2s. 6d. cloth.

Stories of Edward and his little Friends.

With 12 Illustrations. Second Edition. 3s. 6d. plain; 4s. 6d. coloured.

Sunday Lessons for little Children.
By Mrs. BARWELL. Third Edition. 2s. 6d. plain; 3s coloured.

The Grateful Sparrow.
A True Story, with Frontispiece. Fifth Edition. Price 6d. sewed.

BY THE SAME AUTHOR.

How I Became a Governess.
Third Edition. With Frontispiece. Price 2s. cloth, 2s. 6d. gilt edges.

Dicky Birds.
A True Story. Third Edition. With Frontispiece. Price 6d.

My Pretty Puss.
With Frontispiece. Price 6d.

Dissections for Young Children;
In a neat box. Price 5s. each.

1. SCENES FROM THE LIVES OF JOSEPH AND MOSES.
2. SCENES FROM THE HISTORY OF OUR SAVIOUR.
3. OLD MOTHER HUBBARD AND HER DOG.
4. LIFE AND DEATH OF COCK ROBIN.

ONE SHILLING AND SIXPENCE EACH, CLOTH.

TRIMMER'S (MRS.) OLD TESTAMENT LESSONS. With 40 Engravings.

TRIMMER'S (MRS.) NEW TESTAMENT LESSONS. With 40 Engravings.

ONE SHILLING EACH. CLOTH.

THE DAISY, with Thirty Wood Engravings. (1s. 6d. coloured.)
PRINCE LEE BOO.

THE COWSLIP, with Thirty Engravings. (1s. 6d. coloured.)
THE CHILD'S DUTY.

DURABLE BOOKS FOR SUNDAY READING.
Illustrated by J. GILBERT. Printed on linen.
Price 6d. each.

SCENES FROM THE LIVES OF JOSEPH AND MOSES.
SCENES FROM THE LIFE OF OUR SAVIOUR.

DURABLE NURSERY BOOKS,

MOUNTED ON CLOTH WITH COLOURED PLATES,

ONE SHILLING EACH.

1 Alphabet of Goody Two-Shoes.
2 Cinderella.
3 Cock Robin.
4 Courtship of Jenny Wren.
5 Dame Trot and her Cat.
6 History of an Apple Pie.
7 House that Jack built.
8 Little Rhymes for Little Folks.

9 Mother Hubbard.
10 Monkey's Frolic.
11 Old Woman and her Pig.
12 Puss in Boots.
13 Tommy Trip's Museum of Birds, Part I.
14 ———————— Part II.

BY THOMAS DARNELL.

PARSING SIMPLIFIED: An Introduction and Companion to all Grammars; consisting of Short and Easy Rules (with Parsing Lessons to each) whereby young Students may, in a short time, be gradually led through a knowledge of the several Elementary Parts of Speech to a thorough comprehension of the grammatical construction of the most complex sentences of our ordinary Authors, either in Prose or Poetry, by THOMAS DARNELL. Price 1s. cloth.

GEORGE DARNELL'S EDUCATIONAL WORKS.

The attention of all interested in the subject of Education is invited to these Works, now in extensive use throughout the Kingdom, prepared by Mr. George Darnell, a Schoolmaster of many years' experience.

1. COPY BOOKS.—A SHORT AND CERTAIN ROAD TO A GOOD HANDWRITING, gradually advancing from the Simple Stroke to a superior Small-hand.

LARGE POST, Sixteen Numbers, 6d. each.

FOOLSCAP, Twenty Numbers, to which are added Three Supplementary Numbers of Angular Writing for Ladies, and One of Ornamental Hands. Price 3d. each.

   . This series may also be had on very superior paper, marble covers, 4d. each.
   " For teaching writing I would recommend the use of Darnell's Copy Books. I have noticed a marked improvement wherever they have been used."—*Report of Mr. Maye (National Society's Organizer of Schools) to the Worcester Diocesan Board of Education.*

2. GRAMMAR, made intelligible to Children, price 1s. cloth.

3. ARITHMETIC, made intelligible to Children, price 1s. 6d. cloth.
   . Key to Parts 2 and 3, price 1s. cloth.

4. READING, a Short and Certain Road to, price 6d. cloth.

GRIFFITH AND FARRAN, CORNER OF ST. PAUL'S CHURCHYARD.

www.ingramcontent.com/pod-product-compliance
Lightning Source LLC
Chambersburg PA
CBHW021938190326
41519CB00009B/1057